HOPE AND DECEPTION IN CONCEPTION BAY: MERCHANT-SETTLER RELATIONS IN NEWFOUNDLAND, 1785–1855

In late eighteenth- and early nineteenth-century Newfoundland, the evolution to colonial self-government within the empire was accompanied by an economic transition from a migratory to a residential fishery. This was the beginning of the modern liberal order for Newfoundland.

The standard view is that the truck system, wherein merchants supplied fishing families with provisions, gear, and so on against the season's catch, shamefully exploited resident fishermen, as well as planters and servants. Sean Cadigan reviews the economic and social developments of this period from a new perspective. He contends that the persistence of independent commodity production in the fishery of northeast-coast Newfoundland from 1785 to 1855 cannot be attributed to merchant-imposed truck credit practices. He calls for a reassessment of the truck system as a realistic accommodation to the limited possibilities and requirements of the local economy. The rise of the truck system and the household-based fishery was above all a historical outcome which involved the adjustments of settlers, merchants, and governments during a complex period of transition. Elements of the staple model are used to suggest that the resource base of the fishery and the legal institutions of the initial fishing industry limited the ability of fishing families to respond otherwise to exploitation by merchants. Later, reformers struggling for colonial self-government obscured the staple restraints on fishing families in order to discredit fish merchants politically by saying the latter purposefully used truck to impoverish the fishery and prevent agricultural development in order to preserve their hegemony in Newfoundland's economy and society.

Besides newspapers accounts, missionary correspondence, and local government records, Cadigan makes use of court records that have never before been systematically used. These records provide evidence that serves as the basis for his discussion of family production in the fishery, the unsuccessful attempts by families to diversify production through agriculture, the gender division of labour, and economic development.

SEAN T. CADIGAN is a professor in the Department of History, Memorial University of Newfoundland.

$\rightarrow$ 4 strands

① Labour relations
② Did merchants restrain devpt
③ agriculture + diversification
④ Liberalism + reform

The strands are linked together by myth of newf'd
history → poverty of fishing people, failure of
politicians to diversify economy (colonial exploitation)
the merchant oligarchy, the evils of truck,
the heroism of reformers.

There are some peculiarities, a few ambiguities, and even a few points that fail to convince.

SEAN T. CADIGAN

- bibliog/footnotes errors

- neglect of religion

- Agriculture promotes subsistence not commercial

a desire for monocausal explanations { ambiguities about what 'credit' meant + how it was used → sometimes it is presented an enquiry other times geographic determinism

Hope and Deception in Conception Bay: Merchant-Settler Relations in Newfoundland, 1785–1855

These criticisms cumulate in not, in a wider sense of disappointment. Though Cadigan encourages the idea that this is a local history of social relations, it isn't. We are told little about the nature of the merchants work, or gets no real sense of the interactions among people, no idea of the nature of this society or the bonds of kinship, place + religion. More disappointingly one gets no sense of place, of the geographic or relational space in which these people moved.

UNIVERSITY OF TORONTO PRESS
Toronto Buffalo London

© University of Toronto Press Incorporated 1995
Toronto Buffalo London
Printed in Canada

ISBN 0-8020-0469-5 (cloth)
ISBN 0-8020-7568-1 (paper)

Printed on acid-free paper

Canadian Cataloguing in Publication Data

Cadigan, Sean T. (Sean Thomas), 1962-
Hope and deception in Conception Bay : merchant-
settler relations in Newfoundland, 1785-1855

Includes bibliographical references and index.
ISBN 0-8020-0469-5 (bound)   ISBN 0-8020-7568-1 (pbk.)

1. Fish trade - Newfoundland - History.
2. Fisheries - Economic aspects - Newfoundland.
3. Fisheries - Social aspects - Newfoundland.
4. Newfoundland - Economic conditions.
5. Newfoundland - Social conditions.   6. Merchants -
Newfoundland - History.   7. Fishers - Newfoundland -
History.   8. Conception Bay (Nfld.) - Economic
conditions.   I. Title.

HD9464.C33N48 1995   338.3'727'09718   C95-930123-2

University of Toronto Press acknowledges the financial assistance to its
publishing program of the Canada Council and the Ontario Arts Council.

This book has been published with the help of a grant from the Social
Science Federation of Canada, using funds provided by the Social Sciences
and Humanities Research Council of Canada.

Contents

Preface:
The Chimera of Newfoundland
History

Fish merchants are the chimera of Newfoundland history. The fable originates in a Newfoundland historiography which suggested that merchants used what were known as truck practices to manipulate the price of goods purchased from and sold to their fishing clients, in order to insure a favourable total balance of credit and debt at the end of a fishing season. Merchants in the Newfoundland trade also supposedly pressed the imperial government to forbid the development of agriculture, settlement, and settler institutions which might challenge fish-merchant hegemony over the island.[1]

Later dependency and Marxist writers embellished this chimeric fable. For these writers the name of the beast was 'merchant capital.' Gerald Sider, in particular, confused Marx's suggestion that merchant capital was theoretically conservative in the process of capital accumulation, with the historical phenomenon of fish merchants. Sider consequently accepted the earlier view that fish merchants acted to preserve their hegemony by encouraging the state to prohibit agriculture, refuse to recognize property ownership in Newfoundland, and deliberately oppose the formation of a landed gentry. But additionally, merchants impoverished petty producers in the fishery by manipulating wage and credit laws to undercut the growing use of wage labour in fish production. This impoverishment subsequently inhibited the formation of domestic capital by preventing the development of local alternatives to merchant domination through truck, which in turn reduced the amount of cash in domestic circulation, halted the industrial-capitalist reorganization of the fishery, and encouraged 'traditionalism' – an autonomous fishing-village-based, preindustrial, producer culture.[2]

The merchant-capital chimera was a beast of three parts. It supposedly inhibited Newfoundland development through truck, prohibited agriculture, and manipulated wage and credit law. The recent conception of the chimera originates in the transition debates and more particularly in dependency theory and world-systems analyses, which tend to define merchant capital as a force active in the creation of underdevelopment rather than colonial capitalist development.[3] But in the Newfoundland context, this conceptualization of the merchant arises from a history of liberalism's inability to come to terms with a society completely dependent on the fishery.

While the literature which supports the chimera is large, there are a number of reasons why I felt compelled to write a book which dissents from the myth. The first arose from my reading of the established literature on merchants and the development of the Newfoundland fishery, which rejected the simplicity of blaming greedy merchants for exploiting fishing families. This body of work suggests that truck was not something imposed by the venality and avarice of merchants, but rather that it arose as a complex adaptation over time to mutual, if unequal, merchant-fishing-family dependence on salt-cod markets in a region with few other resources to encourage much production outside of the fishery.[4] The second grew from personal experience. Having grown up with the long winters and extremely short summers of Newfoundland, I found it difficult to believe that merchants or anyone else would ever have to worry about prohibiting agriculture here. To be sure, there are scattered pockets suitable for husbandry or root-crop cultivation, but Newfoundland is generally a barren, tundra-dominated island of thin soil, vast rocky barrens, and cold, foggy weather. If people depended on merchants and the fishery, it was because there was little else they could do for themselves.

This is not to say that a better economy and a richer society could not have been founded on the fishery. This book is not an apologia for the merchants, who were interested in little but profiting from the exploitation of those who caught the fish. But this book does reject the ahistoricity of expecting that the merchants should have developed the fishery philanthropically, for the benefit of Newfoundland rather than for themselves. If merchants in other colonial contexts encouraged economic development, it was because they became caught up in complex class struggles which provided lucrative opportunities for profit from the trade of increasingly capitalist local economies.[5]

Merchant capital was profoundly ambiguous about changing productive relationships in colonial societies, often adapting to the opportunities for more diversified local capitalist production rather than trying to hang on to colonial-metropole trade.[6] I wanted to know why class struggle in Newfoundland fishing communities did not produce a local capitalist dynamic which might have attracted merchant attention. Other work dealing with the problem of merchant capital points to the agrarian roots of American capitalism. Particular regions of the North American continent attracted migrants fleeing the expansion of European, especially British, capitalism. These migrants came to America, dispossessed aboriginal peoples, often engaged in petty production primarily for themselves but increasingly for the market, and they did so until the land was full. By then, however, agrarian petty capitalism had in turn created local manufacturing, which grew larger as it fed on the labour of those who found that access to the land was now restricted. (The following introductory chapter will explore this in a more detailed British North American context.)[7]

Newfoundland had almost no petty production in agriculture because its climate and soil would not support it. While staple theory may well be too restrictive for understanding the development of an agrarian society such as Upper Canada, I find it helpful for understanding the contours of the class struggle in Newfoundland.[8] To benefit from a staple perspective, however, does not mean that a historian of class has to fall into the traps of determinism or commodity fetishism by reducing Newfoundland history to a linkage analysis of the cod trade.[9] But as Rosemary Ommer has shown in her history of cod fishing in the Gaspé Peninsula of Quebec, economies dominated by one economic activity such as cod fishing generate little else to help producers escape the merchants' power. Ommer has demonstrated that it was not the cod trade itself which effected Gaspé underdevelopment and prevented fishing people there from seriously challenging Jersey merchants, but rather the lack of much else to work with. The latter used truck and imperial law to organize the fishery so that its wealth and industrial spin-offs benefitted the Jersey metropole. If producers could have found practical means which allowed them to do any number of things – withdraw from the fishery into other production, engage in import substitution, or find willing alternate sources of credit – they might have struggled successfully against the dominance of merchant capital.[10]

I have turned, then, from a staple approach which emphasizes the role of external demand in colonial development, to one which emphasizes that resource endowment, particularly the soil and climatic conditions which might encourage family rather than plantation agriculture, could positively influence the settlers' ability to negotiate their relationships with merchants.[11] While I have emphasized the importance of the staple in understanding Newfoundland's history, I would simply like to suggest here that agriculture is a potent force in all colonial economic and social development, but one almost completely unimportant to an island so dependent on marine resources.[12]

Newfoundland fish merchants may well have been beasts, but they were no special ones. Their relations with fishing families unfolded in an environment with particularly limited resources, which did little to encourage local interest in the political and legal infrastructure of a region and a fishing industry often marked by antagonistic capitalist, colonial, and imperial interests. Merchants did not take any special steps to oppose agriculture or manipulate the legal system to prevent social or economic differentiation, because local commodity producers could find no resource base from which to begin such activity successfully. Fishing families, unable to see any way of producing significant amounts of subsistence or capital goods locally, had no choice but to rely on merchant credit and the purchase of imported goods. The overhead costs of credit, in addition to the fishery's labour requirements and legal infrastructure, ensured that fish producers continued to rely primarily on family labour. Yet Newfoundland was not immune to the demands for constitutional reform which arose in the more developed societies of its British North American neighbours. Facing the reluctance of imperial officials to grant representative or responsible government to a dependency almost completely without an agrarian base, Newfoundland reformers and Liberals by the 1850s had conceived the chimera: that merchants had taken special steps to choke the flower of local agriculture and had stifled every attempt by fishing families to break the grip of merchant capital on their livelihoods.

While pursuing this study I have focused on the society and economy which developed on the northeast coast of Newfoundland, the area most important for the resident fishery. Conception Bay receives the greatest amount of attention because it had some of the most favourable agricultural conditions and was the centre of the early nineteenth-century seal and Labrador fisheries. These fisheries were the ones in which planters made the most use of servants hired on wages.

It is also important that I clarify the terminology appropriate to the description of a society with little urbanization, and dominated by non-industrial capitalist production in its fishery. Northeast Newfoundland was not characterized by distinctions between capitalists and workers. Class relations were defined rather by one's relationship to credit, the crucial means of production in the fishery. The primary class division existed between merchants and settlers, of whom the latter almost invariably made their living in the fishery. Merchants were those who engaged in the fish trade by purchasing and exporting salt codfish, fish oil, seal skins, and seal oil. They purchased these products from fisher-clients whose operations were financed by merchants who advanced supplies on credit. Fish merchants, who generally owned their own seagoing vessels, were able to import goods directly for wholesale and retail in Newfoundland, and to export fish directly to external markets. Many of the merchant firms described in this study, such as Pack, Gosse and Fryer, were partnerships of Newfoundland and British principals; but others, such as the Leamons of Brigus, were the clients of firms such as Bowrings, as the Newfoundland port of St John's grew as a mercantile centre. After 1815 a variety of smaller merchants appeared. These were often planters (also called traders or dealers), who coasted and retailed goods supplied to them by larger merchant firms for fish and perhaps provided shipping and other mercantile services, but did not have the capacity to deal directly with markets external to Newfoundland.

Originally all settlers were known as planters. In the context of colonial history, the term *planter* is in fact something of a misnomer and should not be confused with the great plantation owners of the American South. Although the term may be linked to the proprietary colonists of the seventeenth century, by the late eighteenth century Newfoundland planters were simply settlers engaged in a resident fishery, as opposed to those who migrated annually to Newfoundland to fish and then returned to Great Britain. It was residence and ownership of a dwelling, flakes (wooden platforms built on stilts and covered with boughs on which fish were dried), stages (other platforms which bore sheds and worktables where fish were processed and equipment stored), gardens, and boats, not employment of wage labour, that defined one's status as a planter. Planters were household producers who, unlike other fishermen, possessed all the property and equipment to process fish, though both planters and fishermen relied on family labour and merchant credit for provisions and capital goods to prosecute their fishing voyages.

Planters and fishermen occasionally hired servants, usually young men (almost always called youngsters in the trade) but sometimes women, to assist them. Before 1830, when most northeast-coast communities were in the early stages of permanent settlement, servants usually came from Irish or West Country English rural households. Service in the fishery was usually an apprenticeship that allowed these youths to mature while learning their trade and until they could establish their own fishing households, often by marrying into the households of their masters. After 1805 service became rarer and most servants came from local, less well off fishing families, though they still looked forward to establishing their own households. To add to the confusion, all men who actually fished from boats – whether master, family member, or servant – were often called fishermen, while men, women, and children who salted and dried fish on shore – servant or not – were called the shore crew. Settlers and merchants measured the salt fish they processed and marketed in quintals, each quintal weighing about 112 pounds. While I have tried in the text to identify any other terms which might pose problems for those unfamiliar with Newfoundland history, readers might also enjoy consulting the *Dictionary of Newfoundland English*.[13]

Acknowledgments

This book is based on a dissertation completed at Memorial University of Newfoundland in 1991. I would like to thank my co-supervisors, Greg Kealey and Rosemary Ommer, for their support and encouragement. I often find it hard to see what I do not owe in my work to Greg, who has been advising me since my days as an undergraduate. Rosemary eased my entry into the world of fishing people and merchants by providing a vital link between the disciplines of social and economic history and historical geography. She also introduced me to the world of Atlantic Canada workshops and Atlantic Canadian studies conferences, from which I have learned much. J.K. Hiller, the third member of my dissertation committee, also helped a great deal, as did my examiners, Gerald Panting, John Mannion, and Eric Sager, who provided valuable suggestions concerning the book's revision. I especially appreciate Eric Sager's continuing interest in my work, advising me as he always does from the distant shores of Victoria. Daniel Vickers has also stood by constantly with advice and good humour, and I cannot speak highly enough of the support given to me by the Memorial Department of History and the Maritime Studies Research Unit.

Fellow graduate students Jim Armour and Jeff Webb were important sources of advice and camaraderie, Laura B. Morgan put up with a lot of griping about writing and droning about planters, and Mark Leier deserves special mention for too many things to list.

The preparation of this book was made easier by the careful attention of Gerry Hallowell, Agnes Ambrus, Robert Ferguson, and Patricia Thorvaldson of the University of Toronto Press. Allan Greer and Craig Heron provided advice in their capacity as editors of the Social History of Canada series.

The Social Sciences and Humanities Research Council of Canada provided funding for this project's research. I have received publication assistance through an open competition from the Publication Subvention Board of Memorial University for the preparation of maps.

I would like to thank Gary McManus of Memorial's Cartographic Laboratory for the preparation of maps. Irene Whitfield, managing editor of *Labour/Le Travail*, prepared the tables, and more importantly, with Joan Butler provided much moral support over the past seven years.

An early version of Chapter 2 appeared in Daniel Sansom's *Contested Countryside*, and of Chapter 5 in Colin Howell and Richard Twomey's *Jack Tar in History*. Material in Chapters 3 and 7 appeared in my essay on the staple model in *Acadiensis*. I am grateful for permission from Acadiensis Press to use this material again, although it has been much revised.

I dedicate this book to my mother, Joan, who taught me to appreciate reading and education, and to my father, Thomas, for his curiosity and knowledge of the past, and his love of Newfoundland. They have worked their whole lives so that I could learn.

PART ONE: SETTING AND CONTEXT

Introduction

The absence of agrarian life in Newfoundland before 1855 serves as a useful reminder of the importance of agriculture in the rest of British North America. Agricultural settlement was paramount in the establishment of the colonial state. The demand for constitutional change, which reached its crisis in the 1830s and sparked rebellion in the Canadas, popularly arose from the increasingly different visions of the rural people and the colonial state. The lack of rebellion in the Maritimes and the nature of the reform movement in Prince Edward Island can be partially explained by the different rural history of each. The Red River Settlement and the colonization of Vancouver Island broke the hold of the Hudson's Bay Company over the territories beyond the colonial pale. The relative strength or weakness of agriculture is an important part of the explanation for the dynamic but varied changes in British North America throughout the first half of the nineteenth century. Newfoundland, however, stood apart. The gulf which separated it from the British North American colonies was not only the sea, but also its lack of agricultural resources.

The agrarian potential of British North America was not lost on the British officials who sought a solution to the empire's economic and political problems in the last quarter of the eighteenth century. Industrial-capitalist development at home and liberal revolution abroad demanded action. Not only did the British government face direct revolt in North America, but it feared an agrarian one at home. In the Malthusian fashion of the day, imperial officials saw the need to provide an outlet for Britain's growing population surplus. This perception reinforced the demands of loyalist refugees from the American Revolution, that an appropriate haven be established to the north. Loyalists wanted land and security that would be guaranteed by the local administration through representative institutions. The British government was prepared to let Loyalist and British immigrants have the land they desired, but in a controlled way designed to prevent another republican outburst. British policy was accommodationist: while the Quebec Act of 1775 secured seigneurial tenure and French civil law, later constitutional acts divided Quebec into Upper and Lower Canada, gave them English law and representative institutions, and granted Anglo-American settlers the more familiar English property law. The Canadas joined Nova Scotia, Prince Edward Island, and New Brunswick, which had representative governments by 1758, 1773, and 1780, respectively. However, colonial assemblies did not control land alienation. The British state left that in the hands of its appointed governors. These governors were to use land grants to ensure that the nineteenth-century rural populations of British North America understood that they held land by imperial beneficence, and that an ordered, gentry-dominated social hierarchy would prevail as it had not in the lost colonies to the south.[1]

But if imperial and colonial officials hoped that British North America would be a society founded on deference and patronage based on state-controlled access to land, they were frustrated by the aspirations of the agricultural immigrants. In Upper Canada, after an initial policy allowing free grants of land to Loyalist settlers, governors allotted large tracts of land to a variety of speculators, settlement agents, and government favourites such as prominent merchants; or, with more pointed imperial direction, they reserved much land for the use of Crown and clergy, and later land companies. Although such discriminatory practices undoubtedly forced many aspiring farmers into the labour market, other settlers found a variety of ways to gain access to the land. Many found ways to purchase or lease the land, but others simply ignored the property rights of absentee holders and took the land for their own use by squatting.[2]

The farmer-settlers of Upper Canada were often people of education, property, and capital who had chosen to leave behind the shrinking opportunities of capitalist agriculture in Great Britain or the periodic land shortages which developed in the United States. While these settlers did encounter landholders and merchants, they also found a great deal of available land (sometimes of uncertain or illegal tenure) which could be worked in a seasonal round of some considerable potency for colonial economic development. Every season brought work that resulted in a variety of products for home consumption, local sale, or export. While many Upper Canadian farmers might have aspired to commercial production of wheat, it was never a staple crop.[3] The first settlers had no good access to external markets, and they found that local markets favoured the products of mixed farming.[4]

The seasonal round provided a wide variety of goods for export or local consumption. Much of it was food – flour, meat, fruit, vegetables, and dairy products – but industrial goods – such as tallow, hides, grains for brewing and distilling, and forest products such as timber and potash – were also important. The greatest short-term significance of all of these products was probably the choices they gave settlers in their dealings with merchants. Agricultural settlers, like fishing ones, depended on merchant capital. Once they got land, settlers needed the merchants for credit and market access. But in the context of agrarian life, petty production subordinated merchant capital rather than the other way around.

Staple-trade merchants such as Robert Hamilton dominated the economy of early Upper Canada. Hamilton, who began in the fur trade, used his control over the portage at Niagara to develop a transshipment trade. He solicited military contracts and patrons, through which he eventually secured even more business, as well as large landholdings. Hamilton was, however, only a link in a credit chain which stretched from Quebec to the Upper Canadian settlement frontier. As the fur trade waned he concentrated on supplying credit to settlers. Self-interest was a great motivator; settlement aided by his credit increased the value of his landholdings. But Hamilton profited the most by importing goods from his Quebec suppliers, then providing them on credit to smaller merchants who in turn retailed the goods to the agricultural community. The extension of credit in this manner was a risky business that could not depend on the fortunes of one staple good such as wheat. Hamilton's merchant-clients took what they could get from farmers, and consequently Hamilton himself became an exporter of a wide variety of agricultural produce.[5]

Immigrant farmers did not appreciate Hamilton, but they resented him more as a land speculator than as a merchant. Agriculture loosened the ties of credit between farmers and merchants. A great merchant like Hamilton operated from a 'front' town on the Great Lakes–St Lawrence transportation routes (and there were scores of others, such as the Cartwrights of Kingston, the Stone-McDonalds of Gananoque, the Bethunes of Cobourg, and the Buchanans of Hamilton). But the bulk of their clients were dispersed over an often far-removed hinterland in the 'backcountry,' making it difficult for a small group of merchants to control the economic activity of agricultural producers directly.

Mercantile business was consequently competitive. A variety of country merchants and peddlars all tried to capture the business of the agricultural community, sometimes encroaching on each others' clients to satisfy their own front creditors. To cut down on backwoods competition, front merchants sponsored the establishment of both markets in their towns and roads to bring their customers to them. Some merchants, particularly those in western Upper Canada, began to play suppliers of credit off against each other. For those in York and Hamilton it became a game of balancing Laurentian lines of credit against those of the Hudson Valley. Mercantile rivalry encouraged specialization as a variety of retailers, export-commodity speculators, wholesalers, and financiers began to appear.

Mercantile competition to control the agricultural hinterland of Upper Canada was something farmers could use to their advantage. While import-export merchants needed staple goods they could sell abroad, many local retailers and wholesalers would take any farm produce when they could see an opportunity for trading locally. Not only did this mean that farmers had a competitive market for a variety of products, but their own farms gave them additional power. Farmers could not provide for their own subsistence completely, but at times they could live on their own produce alone, to a much greater extent than could fishing people. Thus, farmers were sometimes able to hold goods until they got the price they wanted, allowing merchants less freedom to dictate the prices of goods they took from farmers in return for credit.

The patriarchal structure of farming households also sheltered much of their production from the direct attention of merchants. An engendered division of labour meant that women largely provided their households' subsistence needs, freeing male labour for the production of agricultural goods for market exchange. Such women's work, whether it was keeping poultry, making butter or cloth, or tending kitchen gardens, proved in time to be capable of providing for much more than their own families' needs. Women traded their surplus production locally, contributing to the growth of a domestic market. Women's production initially remained outside the households' trade with merchants because the latter activity was seen as male-dominated. According to Upper Canadian social convention, production for the household, not the market, was a woman's duty. Local exchanges of women's surplus production nevertheless encouraged the growth of domestic industries, particularly in textiles and clothing, and in poultry and dairy products.[6]

The consequences of all this for the economic development of Upper Canada were profound. A variety of merchants took a wide range of farm products, only some of which could be traded to their wholesale and other credit suppliers to repay debt. Much of this agricultural produce had to be converted into credits in the local market. A retailer, for example, might supply grain to distilleries, breweries, and grist-mills, wood to sawmills, or ash to a local potash boiler. The marketing of agricultural produce not only provided credit to farmers, but con-tributed to the industrial diversification of Upper Canada. Merchants had to take a wide variety of goods if they hoped to secure a return on their credit, and they often had to seek markets for these products among the growing numbers of artisans who came to service the needs of rural society. In return, merchants would have to find markets among farmers for artisans' manufactures. Merchants' coordination of their clients' debts helped facilitate the growth of manufacturing rooted in fertile agrarian soil.

The farm economy nurtured industrial capitalism. By mid-century most Upper Canadian land had come under the plough. Farms ranged in size from large ones, worked increasingly by hired labour in spe-cialized production in the longest-settled areas, to newer, smaller ones established on more marginal land by people who often lived more by supplemental wage labour. An increasingly densely populated, ag-ricultural society could not hire its own labour surplus, but such labour could find employment in related, more urban, manufacturing. The blanketing of the countryside with settlers required road construction and dispersed mill and marketing centres. These in turn provided the loci for the wide variety of craftsmen, professionals, and bureaucrats who provided goods and services that farmers could not provide for themselves. As service and manufacturing centres grew, the flexibility of the family farm meant that farmers could often concentrate on crops and livestock that provided many of the inputs – raw materials such as leather for shoes – required for local manufacturing.

Extensive and intensive settlement over the period, as well as con-tinued immigration, made available an even greater labour surplus to fuel manufacturing. Class differentiation followed in both town and country. Agrarian capitalists emerged especially among those longest-settled land speculators whose larger acreages allowed them to specialize further, and who had the resources to hire the labour of their less fortunate neighbours and immigrants. In the towns, a variety of manufacturing – from the processing of agricultural products for shipment and the construction and clothing trades, to the provision of a wide variety of consumer and capital goods required by rural and growing urban markets – provided the context for the emerging distinction between capitalist and wage labourer. Many craftsmen took advantage of the increasing supplies of inputs and labour to expand the scale of their production, which often involved eliminating some mercantile middlemen. But their industries still required a highly skilled labour force with a fairly limited division of labour. Unskilled workers found work not only on farms but also in the timber industry, and more importantly, on the roads, canals, and later on the railways built to facilitate the agrarian trade.

Upper Canada is an example of agrarian life at its most potent, but similar developments occurred in Lower Canada. Although the subsistence-oriented production of peasant households and the non-economic appropriation of surplus characteristic of seigneurialism dominated Lower Canadian agriculture, there appears to have been little actual difference in farm productivity. In any event, new agricultural immigrants were of British origin, and not all that different from those who went to Upper Canada. While habitant households produced for their own needs first, they marketed surpluses in trade with merchants. Over time the increasing commercialization of the countryside, which accompanied the expansion of country shopkeeping, led more habitants into production for the market. Between the 1790s and 1830s their agriculture was prosperous enough to encourage population development, which in turn pushed the limits of available arable land. By 1830 the good land had run out and new farms established on the outermost limits of the old Laurentian heartland were possible only when farmers combined agriculture with wage labour in the timber trade.

In older, settled regions many farmers did well in both production for local markets and the grain trade. As the wheat economy faltered, Lower Canadian farmers switched to livestock and dairy production for the local market. Such agrarian activity contributed to the development of small service centres and local manufacturing. The surplus labour of the countryside either emigrated or found employment in industrial centres such as Montreal, which developed partially in response to local rural needs but primarily because it was the great industrial metropole of Upper Canadian trade. To a certain extent, then, the agrarian strength of Upper Canada was also Lower Canada's. If Lower Canada did not enjoy similarly widespread development in manufacturing, this was likely the result of its having a more limited reserve of agricultural land, rather than poorer farming practices or exploitative seigneurs. Agriculture nevertheless provided much of the labour, attracted many of the immigrants who ended up as wage labourers, and provided many of the industrial inputs required for the colony's more concentrated, but considerable, manufacture.[7]

The Acadians had long established agriculture as an important part of the economy of the Maritime region. Although the Acadian expulsion may have disrupted local agricultural development, the New England settlers (also called planters) who immigrated after 1758 built on their improvements. Joined later by Loyalist and British settlers, in addition to a variety of large-scale proprietary interests, these agricultural settlers and their descendants comprised the majority of the Maritime population by 1810. These settlers practised mixed agriculture on a scale ranging from small holdings to larger, more commercial ones close to the more urban centres.[8] While the fishery, and later timber and shipping, proved to be the staples of the Maritime economy, even by the mid-nineteenth century agriculture was important to domestic diversification. Predominant mixed agriculture produced root crops, garden vegetables, and small amounts of grains. Livestock was most important, providing food in addition to industrial inputs such as leather, wool, and tallow. While only Prince Edward Island exported more agricultural products than it imported, the agriculture of Nova Scotia, New Brunswick, and even Cape Breton, considered individually, dwarfed that of Newfoundland.[9]

Although dependence on truck marked the lives of fishing people in the Gaspé, agriculture provided other settlers in the Maritime region with alternatives to complete dependence on merchant credit in staple trade. The sheer importance of agriculture in Prince Edward Island meant that people in its fishing and shipbuilding industries would not have to rely on merchants for imports of subsistence goods and important capital goods.[10]

In New Brunswick the increasing capital required by an expanded scale of production and forestry regulation meant that large merchants rather than petty producers dominated the timber industry. But petty agrarian producers could find plenty in the timber economy to ensure that merchants could not control their local economy to the same extent as those in the fishery. The labour intensive nature of the timber trade meant that there were more mouths to feed than most timber merchants could provide for. The trade relied on supplemental local farm production, and this alone discouraged the kind of reliance on merchant capital which prevailed among settlers in Newfoundland. Farmers also provided the draft animals and feed required by the timber trade, and could use seasonal employment there to strengthen their farms.[11]

The timber trade was not the only market for agricultural produce. Farmers in the upper St John River Valley produced for local urban markets as well, and for other settlers who were just beginning to clear land. When market conditions warranted, these farmers produced wheat, but they also adapted to the shifting winds by concentrating on husbandry and more diversified food production. Their agriculture in fact equalled that of many parts of the Laurentian region. Throughout the colony, farmers cultivated a variety of crops, taking advantage of market opportunities where they could and often expanding into fairly commercial enterprises when local resources and markets permitted.[12]

Nova Scotia agriculture similarly ranged from subsistence production to production for export to New Brunswick, Newfoundland, and St Pierre and Miquelon. Animal husbandry formed the basis of commercial agriculture in mainland centres such as Annapolis and St Andrews Townships.[13] Even Cape Breton, dominated as it was by merchant capital in the cod fishery and industrial capital in the coal industry, had agricultural resources which attracted Scottish Highland settlers. First-comers got the best land, fronting on the island's lakes and rivers. About 1,500 farmers established commercial farms which produced livestock and some vegetable goods for Halifax, Newfoundland, and New Brunswick markets. Many more farmers settled on poorer backlands. Like the proprietors of more marginal operations throughout the region, they supported themselves by combining their farming with work in the fishery and later in mining. Some found work in local manufacturing, which grew out of agriculture and thus contributed to regional urban development. In addition to the smaller craft-shops of sadlers, wheelwrights, coopers, carpenters, blacksmiths, weavers, and tailors, Cape Breton had gristmills and sawmills as well as some larger tanning and textile enterprises.[14]

All of these rural people, whether in the Canadas or the Maritimes, looked to their governments to assist them in their endeavour to build prosperous new lives. The state acted directly when it had to, giving free land and assistance early on to Loyalists, for example, or assisting emigration schemes which helped to relieve the home country's burdens. Otherwise the state acted to provide essential infrastructure, such as roads, and basic administration, particularly through the judicial system which set the ground rules for people's social and economic activities. Colonists expected and received English common law, although colonial representative governments were supposed to make their own laws without relying on those of the British parliament after their inception. There has been little systematic study of the workings of the judiciary in British North America, but at least in the case of Upper Canada we know that the courts felt it their duty to adjudicate disputes according to English common law. Judges wanted to make sure that merchants and other business people had certain rules by which they pursued their individual gains.[15]

The courts favoured the interests of great merchants or industrial employers in capital accumulation. But in the area of master-servant law, according to English tradition the courts benefited all employers, from the smallest farmer to the greatest shipbuilder, more than it did servants. Although Upper Canada passed no specific master-servant legislation until 1847, its courts acted at times as if English enactments – these were founded on feudal principles, which made breach of contract by an employee a criminal offence – applied in the colonial context. While conservative in many areas, the British North American courts, as part of an Anglo-American tradition of jurisprudence, proved to be quite creative in ensuring that labourers were legally unequal to their employers.[16]

There appears to have been little agrarian or artisanal protest about the law disadvantaging hired labour, but in other areas the state did not respond so well to the needs of the petty producers or the growing bourgeoisie of British North American towns. While the colonial state administered the land and served an agrarian society, it did not do so with an even hand. In Upper Canada there had always been popular discontent about the manner in which government favourites and merchants could secure large, propitious land grants. However, many immigrants of lesser means and status also benefited initially from government patronage in land grants as long as they supported the existing state.[17] When conflict arose it grew more in response to the manner in which colonial governments sponsored infrastructural developments such as canals, which favoured the great import-export merchants and speculators, and estate holders whose land increased in value with immigration. The settlers wanted such funds expended on roads and other internal improvements which benefited the farming majority. Many farmers were not blind to the manner in which those best patronized by government with land grants also received lucrative government appointments or offices such as justice of the peace, which allowed them to ensure that even local improvements benefited their landholdings the most. In Upper and Lower Canada newer settlers were also frustrated by government preference for land companies, which drove up the price of acquiring land. By the mid-1830s a combination of local and international factors contributed to a severe economic recession, which brought to a head popular demands for constitutional change.

The force of rapid social and economic transformation ensured that rebellion met government intransigence in the Canadas. There, colonial oligarchies surrounding the governors held out against demands for reform by playing on the ethnic issue. In Upper Canada, long a destination for American migrants, reformers could be accused of being sympathetic to disloyal American republicans. In Lower Canada, French-English antagonisms complicated a largely but not completely French Canadian reform assault on mercantile privilege. While some reformers faltered at the accusation of disloyalty and looked for moderate change that would not endanger the British imperial connection, radical Upper Canadian reformers and *patriotes* wanted swifter measures. Their advocacy of republican democracy may well have been primarily intended to break, finally, the oligarchic nature of patronage. But radicals in both colonies knew that if they were to succeed they needed the support of the countryside. Consequently, William Lyon Mackenzie's radicalism espoused an agrarian democracy which promised government policies that would benefit prosperous farmers and the mechanics who serviced their needs. While the *patriote* leaders initially defended seigneurial tenure as an underpinning of French Canadian society, they later began to press for reform of the feudal-based exactions of a system which increasingly burdened a peasantry facing a limited supply of land and economic recession.[18]

The rebellions failed, but they did send a strong signal to imperial authorities that the burgeoning societies of British North America needed more local control over their administration. So suggested the Durham Report, which opened the way for the achievement of responsible government by mid-century. While the land issue was not as important a factor in the Maritimes, popular concerns there over arbitrary and inefficient government by an oligarchy beyond popular influence animated demands for change as well. The more dispersed nature of agricultural land, and the greater importance of enclave staple production of fish and timber, led to more scattered, ethnically and religiously diverse Maritime communities whose needs were often more parochial and easily met by existing representative governments. There was little reason to support radical reform, in any event, in colonies which depended upon imperial trade and required imperial protection on the high seas. With a much less concentrated agricultural population and comparatively little executive worry about the control of agricultural-land alienation, there was much less discontent about government land policies. While concerns over merchant-dominated, oligarchic control of patronage prompted demands for reform – demands which eventually led to responsible government – Maritime reformers also wanted a government more attentive to the economic needs of a domestic economy based on the exchange of locally produced agricultural products and manufactures.[19]

The demand for a fairer and more rational administration of colonial resources grew, even in the Maritimes, in response to the needs of increasing agricultural settlement. Nowhere in the region was the agrarian basis of the reform movement as clear as it was in Prince Edward Island. There, settlers found that most of the island's good land had been taken up by often absent proprietors. Few settlers had access to freehold, but rather faced exacting rents and uncertain ownership of improvements or tenure. But the sheer amount of land held on unimproved speculation by proprietors gave settlers an opportunity to use the relatively non-radical reform process of escheat, whereby the Crown could legally take back proprietors' unsettled holdings (on which they often were not yielding rents to the public coffers) and make them available as freeholds to smaller farmers. While the urban bourgeoisie of Prince Edward Island demanded government reform designed to secure internal economic improvement, they found they could build a broadly based political movement only by becoming allied with the agrarian interest of escheat supporters. An essential part of the quest for liberal-democratic reform on the island was a rural one, which demanded fairer administration of land which would benefit the productive majority of the population who lived by the cultivation of the soil.[20]

The importance of agricultural settlement to the emergence of a liberal democratic order is well illustrated by developments across the continent, in the territories controlled by the Hudson's Bay Company. The growth of white settler colonies at Red River, but more immediately on Vancouver Island, foretold the end of both Hudson's Bay Company suzerainty and aboriginal control of local economic resources. Vancouver Island settlers demanded that company influence over the judiciary and control over land prices be broken. Settlers fought for legal reform and an elected assembly because they saw these as indispensable tools in the fair regulation of access to land, which the Hudson's Bay Company had been alienating in ways that favoured itself and its highest-ranking officials to the detriment of other settlers.[21]

All of these agrarian-rooted developments in the British North American colonies have almost nothing to do with Newfoundland, except to illustrate by contrast. Plates in the second volume of the *Historical Atlas of Canada* starkly illustrate the essential difference between Newfoundland and the other colonies. In 1844 Lower Canada produced millions of bushels of field crops, primarily oats and potatoes, which supported a well-developed animal husbandry. Wheat was still an important crop. While much of Upper Canada's agricultural land was still unoccupied bush in 1851, the colony had thousands of acres of improved land producing millions of bushels of field crops and supporting about 2.5 million head of livestock, mainly sheep, hogs, and cattle. As in the Canadas, agricultural settlement covered the Maritime map. The Maritimes had less land under improvement than the Canadas, but the amount still lay in the millions of acres.

Newfoundland, by contrast, had only about 40,000 acres of land under cultivation in 1857. After about thirty years of intensive, government-sponsored agricultural improvement, the island had only this paltry amount to show for the work of its farmers. Newfoundland agriculture yielded only 0.5 million bushels of potatoes, compared to the millions of bushels of oats, wheat, and potatoes produced in the Maritimes in 1851. One or two counties of Cape Breton alone surpassed Newfoundland's total production of hay, while all of Cape Breton produced about twice as much livestock as Newfoundland.[22] Contrary to one recent assessment, agriculture in Newfoundland was

not at all the important economic activity it was to its colonial neighbours.[23] This is not to say that the case of Newfoundland has no con-
tribution to make to our understanding of the continental experience.
Its history of near-complete dependence on the fishery, and the stunted
domestic development that resulted, reinforced the importance to Canadian development of all agricultural beginnings, in even more marginal areas such as Cape Breton.

Bit of the one-road to devpt model

The rest of this book illustrates the different socioeconomic and
political developments which follow when a colony is founded on a
fishery. Part 1 establishes the political-economic framework for the
rest of the study. It argues that it was the lack of any significant resource base other than the fishery – not merchant hostility – that
was responsible for the slow development of settlements and colonial
political institutions at Newfoundland.

Part 2 comprises three chapters which together establish the manner
in which the dominance of the fishery shaped outport society. Chapter
2 demonstrates that household monostaple production allowed little
class differentiation among fishing people, except during the unusually
favourable market conditions of war. Households continued to rely
on family labour rather than wage labour, even during the war, because
the latter could only be paid for by resorting to merchant credit. Merchants did not deliberately impoverish planters, but as we shall see,
the latter found that their best opportunity for social and economic
advance in northeast Newfoundland lay in assuming mercantile roles
themselves, not in expanding their scale of production through the
hiring of more labour. Chapter 3 establishes that fishing households
depended so much on merchant credit because, despite their best efforts, they could find few outside economic activities which could provide any of the consumer or capital goods they required in order to
live and work. Merchants actually encouraged settler agriculture, as
did the state, but they found that they, like the settlers, were constrained by Newfoundland's resource environment. Chapter 4 suggests
that the pervasiveness of the fishery affected even the nature of relationships within the household. Without the sustenance of agriculture, and harnessed by the labour requirements of the fishery, gender
divisions in northeast society took on a very different aspect than
that of the British North American colonies.

Part 3 examines the relationships between fishing people and merchants.[24] Chapter 5 suggests that the legal administration of master
and servant in the fishery departed from English common law and

legislation. Newfoundland settlers, unlike their colonial neighbours, had to deal with laws designed by imperial officials to ensure that the fishery would be exploited directly for the good of the empire rather than the colony. Merchants and planters alike opposed these laws. While the law of the fishery unintentionally privileged servants over employers, in the long run it reinforced the trend of planter reliance on family labour and encouraged truck credit practices, not only by merchants but by the planters themselves. Chapter 6 examines the paternalistic accommodation of truck itself. Credit was a chain which linked fishing people and merchants. While the pull may have been unequal, a considerable struggle over credit demonstrates that fishing people were able to negotiate the terms of their exploitation to some extent.

Part 4 scrutinizes the central concept of the merchant chimera. Chapter 7 examines the manner in which the small but vocal bourgeoisie of St John's began to rail at their exclusion from patronage, in a way that was superficially similar to the reform movements of British North America. Cognizant of the fact that imperial and local authorities were reluctant to extend even representative government to an island without an agrarian base, Newfoundland reformers argued that the island would have had such a base if it were not for the merchants' supposed opposition to agriculture. While early governors saw through the obfuscation of such reform rhetoric, imperial authorities, tired of relief expenses for a colony with such a restricted economic base, gave in to reformers' wishes in 1832 by granting a legislature, but not responsible government. While reformers continued to agitate on the agriculture issue, they found that the governors increasingly accepted their views. Governors were disenchanted with merchants who aggravated the relief problem by withdrawing credit as international markets for fish worsened.

Chapter 8 examines the manner in which the Newfoundland liberal reform movement steered Newfoundland on a policy course which bore little resemblance to the quest for the kind of liberal-democratic, rational administration of resources that had developed in the Canadas, Prince Edward Island, and Vancouver Island. There was no agrarian population to which such a liberal-democratic message might appeal, but there was a colonial government in Newfoundland desperate to secure agricultural development without being forced to by the emergence of responsible government. Reformers thus abandoned much of the agricultural rhetoric for the cause of wage law. Chief Justice

Henry John Boulton, building on his experience of jurisprudence in Upper Canada, and accepting a series of rulings by previous New-foundland chief justices, tried to bring wage and credit law in the fishery more in line with Anglo-American conventions. Reformers cap-italized on popular resentment against merchants by suggesting that fishing people's problems stemmed, not from actual problems in the fishery but from alleged manipulation of Boulton by merchants looking to undercut the planter fishery. In the end, reformers achieved re-sponsible government by fighting for laws which not only ran contrary to legal developments in other parts of British North America, but also added to the difficulties encountered by planters employing serv-ants. In doing so, reformers distracted settlers from the very real prob-lem of restricted resources – the principal problem underlying their dependence on merchant credit – by encouraging the popular but false belief that merchants dominated the fishery through manipulation of the legal system against fishing people.

[handwritten marginal note: Reformers as villains again → Toryism resurgent in early C19 political history.]

1

Political Economy of the Resident Fishery

The people of scattered fishing communities cleaving to the rocky, near-barren harbours of the northeast coast during the late eighteenth and nineteenth centuries were not likely to challenge the dominance of fish merchants to the same extent as people living in more compactly settled and densely populated areas. The residents of Conception Bay, however, lived in one of the coast's most fertile areas, which also had a more diverse range of fisheries. Their communities were less isolated and more substantial than those of the bays farther up the coast. Recent views that Conception Bay is a largely urban centre neglect the legacy of the region's planter fishery, which developed beyond that of the inshore family fishery. Fishing the northeast-coast waters held in treaty by the French (the French Shore) and Labrador fisheries, along with sealing, allowed some Conception Bay planters to employ more capital and labour than did other fishermen.[1] If we are to understand why planters could not challenge merchant hegemony, then it makes sense to focus on the region in which planters had the best chance of doing so.

Few early nineteenth-century colonial or British administrators acknowledged the uniqueness of Conception Bay. Its fishing communities developed within the same broad political economic framework as the rest of the island, one which did not directly encourage local economic or social development. Newfoundland's settlers worked in a fishing industry initially regulated by policies established by the British Board of Trade and Plantations, which was responsible for imperial development. The Board of Trade long regarded Newfoundland not as an object of settlement but as an industry – the cod fishery – which provided a market for British manufactures and a source of specie through the sale of salt cod in Iberian markets. Although the New-foundland cod fishery was never in fact the nursery for seamen required by the British navy, imperial officials believed that it was. Such beliefs further entrenched resistance by the Board of Trade to any developments which might suggest that the resident fishery was superseding the migratory fishery.[2] Indeed, the Newfoundland migratory cod trade was far more important as a provider of direct employment for the surplus labour of West Country rural, agricultural, artisanal, and labouring households than it was a source of labour for the navy. The migratory fishery also depended completely on the products of West Country manufacturing: clothing, leather goods, foodstuffs, drink, fishing equipment, cordage, and nascent industrial-capitalist shipbuilding and refitting. In addition, West Country merchants dominated the supply of Irish foodstuffs to the Newfoundland fishery.[3]

The migratory fishery did have several disadvantages, which counterbalanced the economic linkages enjoyed by the West Country. Annual trips to Newfoundland caught merchants and fish producers in a cycle of winter refitting of ships and hiring of labour in the West Country, a late March-April sailing for Newfoundland to avoid ice, and arrival at Newfoundland in mid-May with its scramble to find fishing rooms and to repair stages, flakes, and buildings. Only during a much-shortened fishing season from June through August could fishermen actually catch and cure fish and still make the return trip to Europe in September-October, thus avoiding the bad weather of a late-fall Atlantic crossing. Migration in the fishery each season further caused merchants and fishermen to leave their immovable shore-based buildings and stores without security or protection. The transatlantic fishery was also vulnerable to the depredations of England's enemies during the many wars of the eighteenth century.[4]

THE RESIDENT FISHERY

Settlement at Newfoundland became the West Country merchants' solution to the problems of the migratory fishery. As early as the seventeenth century merchants from London and Bristol supported proprietary colony schemes in the belief that a resident fishery at Newfoundland would lengthen the fishing season, cut down on the risks of transatlantic crossings, allow fish to be stored at Newfoundland to await better market conditions in Europe, and encourage fish producers to lower the overhead costs of the fishery by finding some of their own subsistence in local cultivation and timber resources. Proprietary colonies such as John Guy's at Cupids, Conception Bay (established in 1610), provided an important basis for Newfoundland's permanent population. Guy, like other proprietors, found that only the fishery provided profitable commodities for trade. Although West Country merchants proved hostile to Guy's attempt to monopolize the best shore facilities for the fishery, the Cupids colony failed because it cost more than it earned. Proprietors found that their settlers could barely eke out a rough subsistence from the soil, let alone produce anything else besides fish for trade.[5]

The people who remained behind after the failure of Cupids and other proprietary schemes did contribute to the permanent settlement of Newfoundland. These settlers did not all stay at Cupids; by 1675 some had settled at Harbour Grace. As long as the migratory fishery dominated, residents watched their communities grow slowly as people came and went according to the vicissitudes of the fishery. Yet some migrants stayed and communities developed, expanding northward through Trinity and Bonavista Bays to the islands of Fogo and Twillingate in Notre Dame Bay (see map, p. 21). Dorset merchants assisted settlement for self-interested reasons: they wanted settlers to expand into fur trapping and seal hunting – winter and early spring activities – as well as into the inshore fishery. The Seven Years War (1756–63) and the American Revolutionary War (1775–83) seriously disrupted the migratory fishery. Such interruptions further encouraged the resident fishery, despite interwar attempts by government to revitalize the migratory fishery.[6]

Conception Bay was a region of early permanent settlement. The bay had a well-established population by the 1740s, and until the 1770s contained between 35 and 40 per cent of Newfoundland's total population. Most settlement clustered between Carbonear and Harbour Main, an area with better fishing and farming resources than most other parts of the island. With the exception of strong Catholic Irish communities in Harbour Grace, Carbonear, Brigus, and Harbour Main, settlers were mostly Protestant English (see map, p. 23). By 1805 Conception Bay had a much larger population of families of long-standing residence than the settlements at Trinity, Bonavista, and Notre Dame Bays. Settlement in the latter areas followed the earlier pattern established in Conception Bay. With good shelter, shore facilities, seal and salmon resources, water and timber, and harbours close to good fishing grounds, Trinity, Bonavista, Greenspond, Fogo,

Protestants

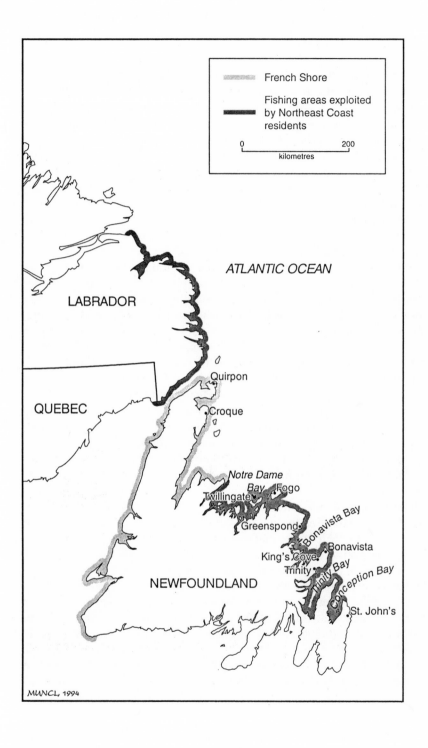

French Shore

Fishing areas exploited
by Northeast Coast
residents

0   200
kilometres

ATLANTIC OCEAN

LABRADOR

QUEBEC

Quirpon

Croque

Notre Dame
Bay   Fogo
Twillingate

Greenspond

Bonavista Bay

Bonavista

King's Cove

Trinity   Trinity Bay

Conception Bay

NEWFOUNDLAND

St. John's

MUNCL, 1994

and Twillingate became the first occupied areas from which settlement spread out into the bays along the northeast coast. British migrants settled first in areas with the best resources, but when settlers' families developed to the point at which they could no longer be supported by such resources, their sons and daughters moved on to other, less well-endowed areas as new immigrants did.

Merchants facilitated settlement expansion. Their premises, established well in advance of community settlement, served as the hub from which the latter grew. By the end of the eighteenth century only Conception Bay families had grown to the point where they could supply planters' labour requirements locally. Some of these people actually began to migrate up the northeast coast as a result of increasing population pressure on local resources, although the growth of the seal and Labrador fisheries allowed Conception Bay to support a larger population than its own resources would otherwise allow.[7] By 1845 more people were leaving Conception Bay than were settling there. The rest of the northeast coast received some of its surplus population until about 1870, when Trinity, Bonavista, and Notre Dame Bays could no longer contain their own natural increase, let alone support new immigrants.[8] The migrations which filled up northeastern bays were family affairs: as first settlements exhausted local resources, family members would branch out to new areas, allowing outsiders to join their settlements only through intermarriage; when an area's resources could no longer support further population growth, families sent their surplus members out to even newer settlements, the remainder forming tightly knit communities which did not allow outsiders access to their resources.[9]

The fishing people who settled the northeast coast discovered that Newfoundland's climate and soil-based resources were not sufficient to support a resident population, let alone an economy much diversified beyond the cod fishery. Late glaciation had left behind a coarse shallow soil and steep-sloped, broken terrain unsuited to agriculture. The small areas of land that did have agricultural potential were scattered widely throughout the Avalon and Bonavista Peninsulas. Even in these places, Newfoundland's extremely variable and harsh weather further restricted agricultural potential.[10] Conception Bay fared better than most areas, however. The northern shore of the bay sheltered the shore south of Carbonear from the chilling effects of prevailing westerly winds. Furthermore, these winds, in their muted form, mitigated the harsher weather which could blow in from the North Atlan-

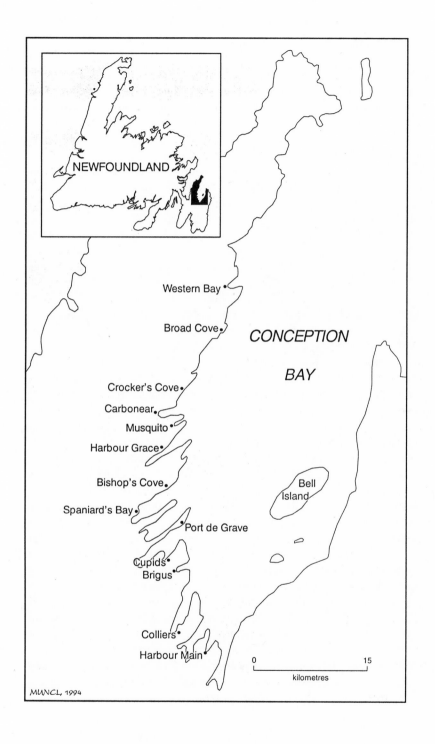

NEWFOUNDLAND

Western Bay •

Broad Cove •

CONCEPTION

BAY

Crocker's Cove •

Carbonear •

Musquito •

Harbour Grace •

Bishop's Cove •

Bell
Island

Spaniard's Bay •

Port de Grave

Cupids •
Brigus •

Colliers •

Harbour Main •

0   15

kilometres

MUNCL, 1994

tic. Finally, the soil and timber of Conception Bay was generally better than that of most other areas of settled Newfoundland.[11]

The American Revolutionary War and the Napoleonic Wars encouraged the development of the resident economy and society in Conception Bay. The two wars generally interfered with British migratory trade, forcing merchants to rely increasingly on resident fish supplies. The loss of food imports from New England after 1775 forced Newfoundland residents to turn to local agriculture, no matter how limited. The successful establishment of American independence restricted the ability of many fishing people to migrate to what had become the United States. Although the migratory fishery began to revive after 1783, war with France in 1793 decisively disrupted British migratory production. The resident fishery again flourished, spurred on by the American Embargo Act of 1807 and the War of 1812, which limited the otherwise formidable competition of the new American codfishery. Meanwhile, the British invasion of Spain opened peninsular markets for Newfoundland fish, and encouraged rising prices.

While wartime inflation afflicted residents, the availability of Irish labour helped to offset planters' wage costs. The Irish had been developing as an important labour supply for the Newfoundland fishery since the 1720s and 1730s, when West Country merchants began to call at Waterford and Cork looking for provisions to trade in the island. The increasing demand of the British war effort on West Country labour augmented the importance of Irish labour to the Newfoundland fishery. The exclusion of ships in the Newfoundland trade from the expensive regulations of the Passenger Act of 1803 further allowed Irish servants with little capital to migrate to Newfoundland.[12]

Both Irish and English settlers became planters, but it was largely the Irish who looked for service in the seal and north shore fisheries of Harbour Grace and Carbonear between 1811 and 1820. Post-war depression hurt these relatively propertyless Irish servants more than it did other Newfoundlanders, making them more likely to follow the growing sectarian-influenced movement for constitutional reform.[13] Philip Henry Gosse (clerk to the Carbonear firm of Slade, Elson & Co.), describing Conception Bay in the early 1830s, suggested that north shore planters in the inshore fishery tended to be Wesleyan Methodists, while the servant population of the Harbour Grace-Carbonear area was dominated by Irish Roman Catholics. Gosse noted that the Irish and English settlers did not get along well together.[14]

The prosperity of the war years encouraged merchants and planters to invest capital in the fishery. Hostilities between Great Britain and France led to the temporary exclusion of the French from the French Shore. By 1798 many fishermen, especially those from Conception Bay, engaged in a migratory fishery aboard schooners between Quirpon and Cape St John. When the end of the war saw fishing rights in this area returned to the French, fishermen from the northeast coast redirected their schooners to the coast of Labrador. By the early 1820s French reassertion of their treaty rights confined Newfoundland schooners to the Labrador fishery alone (see map, p. 21). Floaters – fishermen who lived on board their schooners while fishing – brought their fish back to Conception Bay for processing, while stationers established shore bases from which to conduct an inshore fishery much as they would back home. Through the first half of the nineteenth century the size and importance of the Labrador fishery grew as a supplementary industry to support the increasing population of Conception Bay, although its product was cheaper and inferior in quality to that produced at home. On its own the Labrador fishery was not a viable industry; it existed primarily as a means to employ seal-fishery schooners during the off-season.[15]

The seal fishery attracted the resident population during the years of the Napoleonic Wars. Seals could only be harvested in spring when the large herds drifted southward on ice floes down the northeast coast.[16] The early seal fishery had been a small-scale affair, conducted either from shore or in inshore waters, on the same small craft used in the inshore fishery. However, the harp seals' habit of breeding in large congregations on offshore ice encouraged investment in schooners with which planters and servants could travel to the herds to harvest the white-coated pups, whose hides and fat were most valuable in both quantity and quality shortly after birth. Schooners also allowed residents to go to the herds free from reliance on weather and ocean currents that would bring the ice inshore.[17]

Throughout the first half of the nineteenth century the outfitting of and earnings from the seal fishery served as important stimulants to the economy of northeast-coast communities, especially Conception Bay. The growing seal fishery of the early 1830s employed increasing numbers of ships and men, and the need to keep capital employed year-round ensured the growth of the Labrador fishery as summer employment for the increased number of schooners used by merchants and planters in the seal fishery. Through the early 1840s the seal fishery boomed while the cod fishery lay in depression. After the 1860s, with the advent of costly steam technology, Conception Bay lost its dominion over the seal fishery to St John's merchants with their greater supplies of capital.[18]

There is no reliable data upon which to base a quantitative assessment of social and economic growth in Newfoundland up to 1836. Permanent population growth accelerated after the 1780s, until 1830. Before the war with France in 1793, the permanent population approached 10,000, reached a total of 20,000 by 1815, and 40,000 by 1830. Conception Bay experienced the most rapid growth of any Newfoundland region. By the 1820s its average permanent population was 13,345, or 34 per cent of the island's total population of 39,406. The 10,914 (average) permanent residents of St John's were second, at 28 per cent, while the rest of the northeast coast was third, with 8,996 people (23 per cent). The southern Avalon and southwest coast was last, with 6,151 people (15 per cent). Household production with family labour dominated the longest-settled population of Conception Bay.[19]

Census data collected by the government of Newfoundland after 1832 suggests that the growth of settlement on the northeast coast was based on the expansion of the household fishery and on family labour. In the 1836 census almost no household heads were listed as servants in Conception Bay. Servants constituted approximately 1 per cent of the household heads in Trinity and Bonavista Bays (no return exists for Fogo-Twillingate). Not only did almost no household on the northeast coast survive by wage labour, but there appears to have been no particular concentration of the means of production in the hands of a few. In 1836 Conception Bay averaged four boats for every ten households, although in 1845 and 1857 this number rose to six. In 1836 there was also one fishing boat for every household in Trinity Bay, and one for every two in Bonavista Bay. The 1845 census demonstrates little change, although Fogo-Twillingate averaged only six boats for every ten households. By 1857 there was one boat for every two households in Trinity Bay, about 1.5 for every two in Bonavista Bay, and 2.5 for every two in Fogo-Twillingate. Northeast-coast households each held an average of a half-acre of improved land in 1836. While Conception Bay households managed to add about two acres per household by 1857, households in the other bays averaged only one. The overall picture is one of households almost completely dependent on the fishery, not on wage labour, for their livelihoods, with few agricultural resources to fall back on. People without boats would have to come to some sort of terms with those who did, but

[handwritten margin note: Had servant pop. disappeared?]

the absence of servants as household heads implies that labouring in } *or still migrants*
the households of others was most likely the work of the young, a
stage in life to be passed through before they established households }
of their own.[20]

GOVERNMENT

Settlement based on household production and family labour grew according to the fortunes of war and transatlantic trade. Settlers found direct encouragement at first from neither merchants nor the state. But the West Country merchants' growing reliance on a resident fishery throughout the eighteenth century increasingly came into conflict with both the Board of Trade, which opposed the growth of settlement in Newfoundland, and the merchants, who opposed self-government for the island. Merchants feared that a Newfoundland government might regulate the fishery in favour of residents through trade with the rest of British North America, or (after the American Revolution) with the United States.

The history of Newfoundland's government must be understood within the context of imperial anti-settlement policies and the uneasy interdependence of merchants and settlers. The British government granted the Western Charter to British subjects involved in the Newfoundland fishery in 1634, and confirmed it in 1661. The charter attempted to support both minimal residence and the migratory fishery by limiting the right of residents to enclose land, exploit timber resources, or exclude migratory fishermen from shore facilities. The mayors of the West Country towns and cities involved in Newfoundland trade were responsible for the administration of justice.

Civil war in England ensured that official attempts to limit settlement would not succeed. Many fishermen left their fishing craft at Newfoundland, where they were safe from the depredations of war, and travelled there each summer to catch fish. It was not easy for authorities to distinguish permanent settlers from migratory fishermen who stayed only for the duration of the war. Although the government accepted the merchants' dependence on permanent and semi-permanent residents throughout the late seventeenth century, official hostility to the colonization of Newfoundland continued. Persistent threats from the French over possession of the island served as a counterweight, leading the British government to accept some settlement. The Western Charter failed to prevent residents' attempts to monopolize resources to the detriment of the migratory trade, forcing the British government, in the 1670s, to contemplate disallowing settlement altogether. This new policy would encourage the migratory fishery and prevent Newfoundland from developing along the lines of New England. The British government felt that sea power without settlement was sufficient to hold Newfoundland against the French. A new charter in 1671 tried to end settlement, but authorities did not rigorously enforce it because they were not sure they should, and because they had no way to get rid of settlers who were already there.[21]

Official belief that Newfoundland was valuable only as a fishery and naval nursery, but needed some form of local regulation, led to Parliament passing legislation in 1696 (10 and 11 Wm. III c. 25). This act forbade planters from taking as their own any rooms not occupied before 1685. But by ratifying property held up to 1685, and not forbidding planters to improve and extend such property, this legislation indirectly encouraged settlers. Fishing admirals, the first ships' captains to arrive in a port for the fishing season, were confirmed in the informal administrative and limited judicial authority they had previously exercised in the fishery. The commanders of the naval convoys became in turn an appellant authority over that of the fishing admirals.[22]

West Country merchants came to rely more heavily on supplying the resident fishery, and they sent out their own agents to reside in Newfoundland. Merchants and planters there faced problems of enforcing agreements over credit. By the early eighteenth century merchants began to suggest that the British government appoint magistrates with a limited authority to regulate the fishery. To counter the lack-lustre efforts of the fishing admirals, commanders of the naval convoys appointed justices of the peace for the winter of 1728, and by 1730 allowed them to sit during the summer to decide civil matters. Merchants, whose agents dominated this magistracy, came to accept such limited civil authority, and the Board of Trade turned a blind eye to it. British authorities insisted that criminal offenses be tried in England, but the expense of this eventually led the British government to establish a Court of Oyer and Terminer at Newfoundland by an Order in Council in 1750.[23]

Growth in executive authority slowly accompanied the development of judicial authority. Law and custom prohibited taxation in the fishery, discouraging the British government's interest in appointing a civil governor and administration which would have to be paid for out of the imperial treasury. A 1729 Order in Council commissioned a convoy mander with full civil and military authority over Newfoundland, recognizing his right to delegate judicial authority to magistrates during the winter. The commanders were never given authority to allow a year-round magistracy, although their subordinate officers could hold court in outports as surrogates of the commanders' judicial authority. British authorities expected convoy commanders to counterbalance the arbitrary authority of the fishing admirals, and gradually, after 1729, they included in commanders' commissions a vague mandate to enforce British policy. Smuggling, property ownership, and emigration all proved to be problems which demanded some greater local government.[24] The commander of the British naval squadron at Newfoundland in effect became governor during a period of duty of three years, as set by his superiors in London.

Problems with the regulation of credit relationships between merchants, planters, and servants persisted, particularly as they affected labour discipline. Without local government, apart from the merchant-dominated magistracy, there was little effective supervision of the relationships between the three parties in the fishery. Merchants advanced credit to planters for the provisions and capital equipment they needed to begin the season. If catches or prices were poor, a planter might be tempted to sell his fish to another merchant should he be offered slightly better prices than those of his own merchant. To ensure a return on their credit, merchants would have to take their planters' fish quickly if they thought this situation might unfold. If merchants seized planters' fish, servants would no longer work because they had no hope of being paid at the fishing season's end. Moreover, unpaid servants possessed no means by which they could return home, and imperial policy could not tolerate this threat to a well-trained supply of British seamen who were also consumers of British-made goods.[25]

The American Revolution, while hastening the decline of the migratory fishery, entrenched official opposition to the granting of civil government to Newfoundland for fear that the island might develop its resources and trade in its own interest rather than that of the Empire, as New England had. Yet something had to be done to bring order to the relationships between merchants, planters, and servants. Officials such as Governor Hugh Palliser worried about the manner in which merchants used truck in wage payment to ensure that servants in the fishing industry would have little or nothing left at the end of a season once they balanced their 'pay' against goods and equipment taken on credit. They feared the practice would strand servants in Newfoundland without means to purchase passage home, thus contributing to the growth of settlement. The Board of Trade accepted the recommendations of Palliser, and passed the act which bore his name in 1775. British officials hoped that Palliser's Act (15 Geo. III, c. 31) would revive the migratory fishery, thereby removing the necessity for a government at Newfoundland, guaranteeing that the fishery remain an important market for British manufactures, and supposedly preserving a supply of seamen for the navy. The act consequently focused on protecting servants from the rapacity of the Newfoundland fishery's credit system by articulating two principles which should govern a migratory ship fishery: enforced payment of wages by any merchant who might seize a planter's fish; and the secured return to Great Britain of seamen and fishermen employed in the fishery.[26]

The Board of Trade designed Palliser's Act to protect the wages of British servants temporarily engaged in the Newfoundland fishery. The act specifically provided that anyone employing seamen or fishermen in the Newfoundland trade must agree to set wages in a written contract with their servants before the fishing season began, reserving up to forty shillings of a servant's wages for return passage. While at Newfoundland, servants were not to receive more than half their wages in goods, liquor, or money, and the remainder was to be paid in good bills of exchange drawn on British merchants when the servants returned home. Employers could not avoid paying wages unless they could prove wilful negligence by servants. To do this, merchants or planters had to produce the written agreement in court, and then prove that a servant's actions violated the contract. Servants could only be fined two days' wages for every one day of neglected work. Palliser's Act secured servants' wages by giving them a preferred lien against fish they caught, whether they lay in planters' or merchants' hands. Planters could not profit from the fishery until they had paid the servants' wages, and merchants had to pay planters' servants before receiving any payment for credit extended during the fishing season.[27]

Government in Newfoundland, by 1775, consisted of a limited, naval-based summer government supplemented by a year-round magistracy whose sole purpose, from an imperial point of view, was to buttress the migratory fishery and the empire's interests. The island's isolated position in the North Atlantic, its inhospitable climate, and its bleak landscape meant that imperial authorities could find no reason to apply to Newfoundland any of the forms of colonial self-government

Heavy regulations denote severe problems

which had developed in other British colonies in North America. Palliser's Act increased the power of the naval governor and his surrogates to enforce British regulation of the fishery. It gave them, not the fishing admirals, police and judicial power to issue arrest warrants. The special jurisdiction of the Court of Session and the Vice-Admiralty were wage disputes; the British Admiralty Court or the Privy Council heard appeals. Palliser's Act left otherwise undisturbed the courts of civil jurisdiction which had emerged since 1728.

The courts exercised summary justice through the 1780s. Surrogates and magistrates were accessible and often resorted to by all of the people involved in the Newfoundland fishery. Yet popular discontent with this limited administration escalated. By 1783 the fishery was primarily Newfoundland-based. The increasing residence of West Country merchants or their agents, and the growing complexity of their trade and credit relationships with fishermen, meant that people began to challenge the courts' jurisdiction. Imperial officials responded by creating a Court of Civil Judicature, with full authority in matters of debt, personal property contracts, other property disputes, and wage disputes. Still opposed to colonial self-government, the new court, enacted by the temporary Judicature Act of 1791 (31 Geo. III, c. 29 in 1791; re-enacted as 32 Geo. III, c. 46 in 1792), was to continue the policies of King William's and Palliser's Acts.

This new Court of Civil Judicature, constituting a Supreme Court at St John's, gained jurisdiction over criminal matters and served as appellant authority to the still extant Surrogates' Court. The Judicature Act of 1791–92 provided the first clear means of settling debts, for it recognized the primacy of the wage lien embedded in Palliser's Act (see appendix A), the secondary claims of creditors for the immediate fishing year, and finally the claims of all other creditors. The Supreme and Surrogate Courts exercised all authority in civil matters; the Admiralty Court, while retaining its rule over maritime affairs, no longer had the right to hear disputes involving seamen's wages. The Judicature Act of 1792, like its predecessor in 1791, was the product of a temporary enactment which had to be renewed annually, though without change. The Judicature Act did not become permanent until 1809, when it was re-enacted by Parliament (as 49 Geo. III, c. 27). Aside from the courts' authority, Newfoundland was governed in these years by royal prerogative through naval governors' proclamations. Not until 1818 did the governors stay year-round.[28]

Agitation for government institutions grew in St John's during the

Napoleonic era. The emergence of a resident society and economy, complete with a colonial metropole in St John's, meant that a local bourgeoisie was no longer satisfied to be taxed and regulated from London primarily for the benefit of the empire. Especially vocal in the demand for reform was Dr William Carson, who went much further by beginning to denounce the authority of the naval governors and their surrogates as arbitrary and ignorant, as well as to claim that imperial authority opposed the development of settled agriculture. The remedy, Carson argued, was a resident civil governor and legislature for Newfoundland. Economic depression after 1815 added further vigour to Carson's demands that the British parliament give Newfoundland a colonial constitution, with the same government institutions as the British one. The 1820 whipping by surrogates of two Conception Bay planters, Philip Butler and James Lundrigan, as punishment in debt cases, provided a rallying cry for Carson's fight for colonial self-government in opposition to imperial naval authority. The reformer began to demand judicial reform under the supervision of a local legislature.[29]

In 1820 Patrick Morris, Carson, and other reformers formed a committee of St John's inhabitants who complained to Governor Hamilton about the arbitrary judicial authority of the day, the injustice of taxation without representation, and the lack of a local legislature. Morris represented the growing Irish constituency of St John's, as merchants turned away from what they felt was an excessive demand for colonial self-government. Hamilton forwarded the St John's petition to the Colonial Office. Debate about the petition in Parliament led to another reform petition in 1822, and in 1823 Morris chaired yet another committee which demanded government reform. Reform efforts resulted in a new Judicature Act in 1824, which replaced that passed in 1791. The British government, recognizing that Newfoundland was in fact a settled colony, replaced the Surrogate Courts with circuit courts presided over by magistrates under the authority of civilian judges of the Supreme Court. Newfoundland also acquired a civilian governor with the power to alienate crown land for agricultural purposes.[30]

The Judicature Act of 1824 did not give Newfoundland a legislature. Rather, the British government appointed an advisory council to accompany it, creating a system like that developed for the colony at New South Wales. Its governor, Sir Thomas Cochrane, arrived in 1825 to effect the act, and in 1826 the new circuit courts began their jurisdiction. The governor's council consisted of three Supreme Court judges and the St John's garrison commander. The governor retained full executive authority of the council, which had only an advisory function. An opponent of reformers' demands for a legislature, Cochrane failed to convince the British government to support him. Under pressure from Newfoundland reformers, and influenced by British liberal sentiment which favoured greater colonial self-government through representative institutions, the Colonial Office instructed Cochrane to create a legislature in 1832. This legislature was to be bicameral, consisting of an elected Lower House and an appointed seven-member council with legislative and executive powers. Members of this combined Executive and Legislative Council were usually imperially appointed officials with potentially conflicting duties. The chief justice, for example, joined the attorney-general, colonial secretary, customs collector, and commander of the garrison on the Executive Council, serving as its president. The governor retained the right to adjourn, prorogue, and dissolve the legislature.[31]

The constitution of 1832 persisted – except during the years of the amalgamated legislature from 1842–8, when an equal number of elected and appointed legislators sat in one House – until Newfoundland gained responsible government in 1855. The years between representative government and responsible government were ones of rivalry between a Conservative party, which coalesced around mercantile and Protestant hostility to further reform (which might undermine their monopoly on government patronage), and a Liberal party, which grew out of the reform movement and advanced an eclectic demand for some form of constitutional change which would secure greater patronage for Roman Catholics, particularly those who were members of the growing St John's-dominated Newfoundland bourgeoisie. In 1855, battles between the two groups over a host of issues, many of them sectarian in nature, led the British government to assent finally to self-government for Newfoundland, with an executive authority responsible to an elected House.[32]

Hence closing date

Permanent settlement grew slowly but steadily on the northeast coast throughout the eighteenth and early nineteenth centuries. Settlers began their lives on the coast amid the often contradictory policies and interests of government and merchants, but there is not much reason to believe that such contrariness actually inhibited permanent residence. Fish merchants did come to depend on the resident fishery because of its relative invulnerability to disruption by war, its longer season, and its lower direct costs. If imperial authorities were reluctant to grant it colonial institutions, it was because they saw little more than a fishing industry developing at Newfoundland; neither was it British practice to grant governments to industries. Only with the final disruption of the migratory fishery during the Napoleonic Wars did imperial authorities feel that Newfoundland required a developed civil establishment and colonial status.

PART TWO: THE HOUSEHOLD FISHERY

2

Fishing Households and Family Labour

Merchants had little to do with imposing household production through family labour on Newfoundland fishing people, at least not directly. Merchants never faced a challenge to their domination of the staple trade by a rising planter class which employed wage labour in large-scale production. The planters themselves simply chose to use such labour in an effort to adapt to the capital requirements of producing salt fish. Little evidence exists to support the view of planters as nascent industrial producers except during the unusual economic conditions created by the Napoleonic Wars. Planters otherwise relied on household labour and merchant credit to produce salt cod. While it is true that access to the lucrative north shore fishery – access created by the disruption of French treaty rights to the French Shore – temporarily encouraged many planters to expand their scale of operations through the use of wage labour on schooners. But by war's end in 1815, the restoration of treaty rights and American competition in the fish trade had ended the good market conditions which had additionally supported the planters' expanded enterprises. Left with a much poorer Labrador fishery, planters for the most part retreated into household production.

The almost sluggish pace of settlement in the late seventeenth and eighteenth centuries meant that early planters had to rely on hired labour. Merchants supplied planters with servants initially recruited from the English West Country: usually the unemployed; sometimes artisans, labourers, or orphaned youth of farmers. These young people agreed to join the domestic establishments of the planters for two summers' and one winter's work in exchange for a wage. And it was not unusual for such employment to end with servants marrying into the families of their masters. Servants did not constitute a class in themselves, but rather hired out into what was essentially a transitional stage between the households of their parents and the households they would establish on their own, having become the in-laws of others. In so doing, West Country servants, like the Irish later, found that their migration for work had become transatlantic emigration.

By the second decade of the nineteenth century local households in Conception Bay could supply enough of their own family labour, and the practice of recruiting servants from abroad declined. Instead, planters used hired labour to supplement not supplant that of their families. Merchants supplied servants to their clients as they would any other good or service: on credit. Wages owed by planters were simply another debt on the merchants' books, no more impervious to price manipulation than any other charge. Paying wages meant depending more on merchants, something planters tried to avoid.[1]

Even as early as the late eighteenth century, it was very much the trend for planters to rely on family labour. In 1791 Chief Justice Reeves described Newfoundland planters as being often 'no more than Common Fishermen,' holding little property, completely dependent on merchant credit, and vulnerable to failure: 'one or two successful seasons may possibly work such a man into a little property in his Boat, & Craft; but should one successful season throw him in arrear to his Merchant there is scarce chance of recovering.'[2] War with France later in the decade meant that some planters prospered beyond the state described by Reeves. The absence of French and later American competition in the fish trade led to higher prices for planters' fish. Planters could expand their scale of production in response to these prices by employing Irish labour made available by worsening economic conditions in Ireland.[3] In 1798 Governor Waldegrave could write about a new type of 'Planter who labours for himself without the assistance of the Merchant.' Although few in number, these planters did hire servants to conduct their fishery.[4]

Depression in the cod-fishery at the start of the Napoleonic Wars occasionally created opportunities for some planters to expand through diversification. Without enough earnings to survive in Newfoundland from the fishery alone, many planters turned to sealing, trapping, shipbuilding, and logging to supplement the cod-fishery. Combining sealing and the fishery meant that some planters could obtain enough credit to outfit a schooner, with which they might clear themselves in a year or two of any credit obligations to particular merchants,

and begin to trade independently.[5] While many planters continued to
rely on family production or limited partnerships on a share system,
a considerable number were able to hire servants on wages. Yet even
planters who used wage labour had to deal with local merchants in
truck, earning little above the costs of their fishery and their subsistence. Wartime inflation hurt many planters, who could not obtain
high enough prices for their fish to compensate for high wage rates
and high credit prices for equipment and provisions. The vagaries of
wartime economic conditions could destroy as well as ensure a planter's
independence: merchants' suits against planters for bad debts increased
in the Newfoundland Supreme Court as the war years brought increased prices for imports to the island.[6]

Differentiation among planters thrived on a new enterprise created
by the Napoleonic Wars: the north shore fishery. The north shore
lay between Quirpon and Cape St John, on the westernmost tip of
Notre Dame Bay, and had been held by the French since 1713 as part
of their treaty rights to the French Shore. Although the French continued to claim the north shore as part of their treaty area until 1904,
even when the boundaries of the French Shore shifted in 1783, hostilities with the British from 1793 to 1815 disrupted the French fishery
there. Some planters who had previously been simply year-round
settlers on the northeast coast began to expand their scale of operations
by employing servants on fishing trips to the north shore. This north
shore fishery proved superior to the Labrador waters resorted to by
some planters since the 1760s.[7] Erasmus Gower, Governor of Newfoundland from 1804 to 1807, described the north shore fishery in
such a way as to suggest that wage labour did not supplant family
labour, but rather acted as an appurtenance to it. Planters from Conception Bay hired passage for their families on schooners to get to
the north shore, where men caught fish and their relatives cured it
just as they did at home.[8] Over 100 schooners sailed to the north
shore by 1812, each employing an average of twenty hired servants.[9]

The fact that planters used servants to crew a schooner does not
necessarily mean that they employed servants actually to catch or cure
fish. Planters could well have been coasters – traders who were not
primarily involved in production but who sailed the Labrador and north
shores searching for cargo. The coasters would freight the fish caught
by fishing people they had carried to these coasts in the spring. In
1808, for example, planter Richard Kain sued another planter, Francis
Pike, for £124 damages to fish improperly handled by Pike's schooner

crew as they carried it from Kain's fishing room at Goose Cove on the French Shore to Harbour Grace. Kain proved to a jury that Pike's crew had allowed 197.5 out of 300 quintals of fish to become wet in shipment, damaging their cure. Pike clearly employed labour in this instance, but not in the fishery itself.[10] In a similar case that same year Michael Kain sued planter William Peddle for £100 damages for failing to deliver supplies to him on the French Shore as they had earlier agreed. Again, Peddle acted as a shipper not as a producer of salt cod.[11]

The planters' evolving role in schooner shipping may have been promoting the expansion of their scale of producing salt fish. Francis Pike, in partnership with his mother, Elizabeth Pike (the wife of a deceased merchant), had by 1808 begun contracting the curing of fish caught on the French Shore by Conception Bay planters. Such contracts to process salt fish appear to have arisen from the Pikes' agreements to ship fish for north shore planters. Evidence of this can be found in Robert Ash's suit against Pike for improperly curing his 'trip' of green fish. Ash used a schooner to catch fish on the French Shore, sending two cargoes of green fish to Elizabeth Pike's stages during the 1807 season. Testimony by servants of both Pike and Ash indicate that the former hired shoremen to cure Ash's two loads of fish, which he sent to Carbonear from the French Shore.[12]

The Pikes' operations suggest that the French Shore extension of the northeast-coast planter fishery was clearly leading to some local market diversification and specialization. Owners of capital – both Ash and Pike – employed servants in a manner that little suggests a household relationship. There was regional specialization of labour: planters could concentrate on catching and splitting fish at the French Shore, while all the curing was done at the marketing site in Conception Bay. Planters on the north shore enjoyed a longer season than those at Labrador, and they had to invest much less capital in preparing for a voyage because the shorter trip required fewer provisions.[13]

Planter employment of labour in the north shore fishery ebbed with the tide of war. Governor Keats, in 1814, warned the Colonial Office that prosperity would not last when markets for fish returned to normal, and when planters could no longer afford to pay high wages.[14] British peace negotiations threatened planters in the north shore fishery by readmitting both French claims to the north shore and American fishing rights at Labrador. French and American competition, along with the loss of preferences for British products on the Iberian Peninsula, ended the unusual demand for Newfoundland fish. The resulting fall in fish prices meant less demand for servants above the requirements of the planters' household fishery. Merchants supplying the settlements of the northeast coast, quickly perceiving this, requested that Governor Pickmore compel Newfoundland's surplus labour to leave the island when depression hit the fish trade in 1816–17. Pickmore replied that he had no means to do so, but acknowledged that low prices for their catches forced planters to let go their servants, causing a severe unemployment problem.[15]

The post-1815 depression of the fish trade occasioned some interesting commentary on the household nature of production in the Newfoundland fishery. J. Newart, who described himself as a long-time resident of Newfoundland, suggested that planters were mostly ex-servants, or the descendants of servants who had managed to acquire enough capital in partnerships of two or three to purchase boats to begin fishing on their own account. To be more accurate, planters were those who dried not only their own fish but also, with their families and servants, dried the fish of fishermen without flakes or stages. (This description seems to fit the operations of Elizabeth Pike well.) Merchants dealt with planters through the price-fixing manipulations of truck: they did not tell planters how much they would be charged for provisions and equipment until the merchants knew how much the fish and oil would fetch in the marketplace. Late-war prosperity led merchants to advance more credit to some planters so that the latter could extend their scale of production, but post-war recession ended such expansion. Planters with extensive investment in the fishery fell into insolvency, leaving behind only the family fishery.[16]

In the most developed parts of Conception Bay, around Carbonear and Harbour Grace, families retreated behind the labour of their households so that they might avoid merchant credit as much as possible.[17] The observations of a number of circuit missionaries travelling throughout the northeastern bays confirmed that the inshore fishery survived the 1816–17 depression through the use of household labour.[18] The depression most dramatically affected the immigration of Irish servants. In 1815 about 6.5 thousand Irish passengers, mostly young males, had come to seek work at Newfoundland. Just one year later that number dropped to about three thousand, while in 1817 only 500 Irish immigrants came to the island. Such scant numbers were still larger than the number of other British arrivals. While about a thousand Irish migrants and a handful of others came to Newfoundland every year until 1850, never again was the island to see the massive migration of labour which accompanied the boom of the Napoleonic era.[19]

Fishing people coped with the depression in a number of ways. Planters for the most part became simple boat owners who relied on partnerships on shares with their more impoverished fellows if their own families could not supply enough labour. Poorer planters would work for half the catch of their partners, instead of for wages. During the post-war depression such indigents had all their boats and property seized by merchants when their accounts fell in arrears. The share system, observed Surrogate Judge Captain Nicholas in 1820, proved to be a way in which the insolvent could provide for their families and the solvent could avoid paying wages. Planters of any means had to find any way they could to avoid further indebtedness through truck. Merchants continued to supply them and their servants with goods and equipment, but would not settle prices until the end of the season when the fish came to their stores and the merchants knew what it would bring in the market. Planters could not control this method of pricing but they could control the amount they took, and minimizing the number of servants hired was one way of doing so.[20] Those who could not find a share in a fishing voyage had to resort to other solutions. By 1821 Chief Justice Forbes reported that unemployed servants were leaving the outports to seek work in St John's or to find a way out of the island.[21]

By the mid-1820s planters had generally returned to their status as household producers, or at best, middlemen between fishermen and merchants. The use of labour hired on wages, not shares, had all but disappeared on the northeast coast amid chronic mass unemployment and food shortages caused by persistently low fish prices and the resulting restriction of merchant credit. Post-war depression had eroded the planters' position, forcing them to retreat from the use of hired labour if they were lucky enough to escape insolvency.[22]

The Labrador fishery, still resorted to by schooners from Conception Bay, did not support differentiation among planters as had the north shore fishery. Fishing families from Conception Bay continued an annual migration to supplement their inshore fishery with the catch from the waters of Labrador. Planters in the Labrador fishery used their schooner crews to ship families to the coast in a seasonal round of household activity. The sealing voyage in which the schooners were first engaged before they went to Labrador did not much alter the character of the family fishery, even though it required large numbers of servants. Again, such servants were the young sons of fishing families looking to earn money for their families, or perhaps to start up their own households. After the spring seal fishery ended, these young men either returned to the Labrador fishery with their families or stayed and fished inshore along the northeast coast.[23]

Court records reveal that some Conception Bay planters did use servants to prosecute the Labrador fishery. In an 1817 petition to surrogate Captain Thomas Toker for confirmation of his right to a Labrador fishing room, for example, William Taylor stated that he used one schooner and employed thirteen 'hands' in his fishery there. If his room was protected Taylor planned to use an additional schooner and seven hands. Minutes of other court cases have incidentally revealed that planters employed a number of servants in their schooners to catch fish at Labrador throughout the 1830s. Planters who continued to use servants in their Labrador fishery usually hired them on shares. One such servant, Patrick Rogers, for example, agreed to take a share of fish in return for serving Nicholas Furlong and John Brine at Labrador as a fish splitter during the 1827 fishing season.[24]

The Labrador fishery alone could not sustain planters who had expanded their scale of operations in the north shore fishery. The St John's Chamber of Commerce reported that all Newfoundland fishermen found the Labrador coast's shorter season and poorer curing conditions no substitute for the north shore's resources; the Labrador coast also produced a poorer-quality fish which fetched lower prices. The Newfoundland House of Assembly reported in 1834 that planters relied on supplying goods and services to families they transported to Labrador each year, while they withdrew from production because of the short season, small fish, and poor curing conditions of the Labrador fishery.[25]

Planters who tried to survive in the Labrador fishery after being excluded from the French Shore often supplemented their voyages by coasting the French Shore to plunder on the way home.[26] Some of these raids became the subject of trials in Conception Bay courts. In 1833 James Hope of Carbonear, hired by a French fishing captain, De-'lome, to take care of his property at Croque, pleaded to the Northern Circuit Court that a Carbonear schooner crew raided his premises in October. The court ordered the arrest of the fishermen. In 1840 merchant Thomas Godden complained that the crew of his schooner, led by their master John Sparks, raided Quirpon during the spring.[27]

Prospects were dim for the development of a capitalist production organization in the Labrador fishery. This fishery was only one part of a delicate balance of fisheries in which too much could go wrong. For example, in 1833 Thomas Danson, a justice of the peace at Harbour Grace, reported that shortfalls in fish catches would not allow the production of enough salt fish or cod oil to pay servants' wages. Low catches exacerbated a poor season characterized by bad seal and inshore fisheries. Consequently, merchants only reluctantly advanced credit to planters.[28]

Seven insolvencies involving planters in the Labrador fishery – these surfaced in a sample of writs issued by the Northern Circuit Court (see appendix B) – speak volumes about the nature and pitfalls of capital accumulation in that fishery. Five of the planters appear to have been concerned largely with the actual production of salt cod and oil as the mainstay of their capital accumulation. Planters such as John Long of Port de Grave actually possessed little capital in property to weigh against the credit they took from merchants. Long had a fishing room and equipment – barrels, salt, skiffs, and small utensils – to a value of £33 to balance against debts of over £124 to Martin & Jacob, Robert Prowse, and H. & R.J. Pinsent in 1833. Planter Richard Taylor of Carbonear, in a similar example in 1834, could only balance his assets – a £60 fishing room at Labrador, a farm, and equipment – against about £533 he owed his supplying merchants, Slade, Elson & Co. The large part of Taylor's capital was the credit he had obtained from his merchant.

Other planters, such as John Shea and William Thistle of Harbour Grace (who owed £262 to various merchants in 1837), could not escape dependence on merchants for the credit they needed to employ labour in the Labrador fishery. Servant Laurence Shea's suit against John Shea for the payment of £19 wages occasioned the latter's insolvency. Thistle became insolvent because he could not 'make' enough fish at Labrador to meet the credit he took from Thomas Ridley & Co. for supplies and servants' wages in 1837. Thistle could return only £144 worth of fish and oil against £230 in credit, of which he used £100 to pay wages to six servants, three of whom were probably his sons, David, Thomas, and John. A previous outstanding balance of £260 to Slade, Elson & Co. exacerbated Thistle's troubles: he had only £63 in assets to balance against his debts. It appears that in the cases of both Thistle and Shea, the Labrador fishery could not sustain a constant outlay of capital by merchants to support large-scale production relying on hired labour.[29]

Some planters backed away from relying solely on fish production to accumulate capital by assuming mercantile roles in the Labrador fishery, although this also was no guarantee of success. The 1837 insolvency of Simon Levi, a Carbonear planter, is a case in point. Levi held accounts with approximately 660 people for a total amount of £429. He had begun a small supply business at Carbonear, but continued to operate a Labrador fishery. By 1837 he had managed to accumulate debts of £3,393 to British and Newfoundland creditors, including Conception Bay merchants Pack, Gosse and Fryer, Thomas Chancey & Co., William Bemmister & Co., and Slade, Biddle, & Co. To set against this debt, Levi had only £184 in shop inventory, £10 in two fishing rooms at Labrador, £50 in two plantations at Carbonear, £230 in a half-ownership of the brig *Elizabeth*, £30 in two oil vats, £240 in debts still due him, and £30 in property and furniture, for a total of £774. Simon Levi's estate owed £2,619.[30]

The inventory of the insolvent estate of planter John Meaney of Carbonear in 1843 indicates a similar diversification from the Labrador fishery into mercantile activity (see table 1). Meaney carried a large debt with merchants Pack, Gosse and Fryer, but not only did he own a fishing room at Labrador, he was also a creditor for smaller sums to a large number of other people.[31] Edward Shannahan's debt of £48, owed to Thorne, Hoope and Co. from 1832 to 1836, led to a petition by the planter which stated the problems of using hired servants in the precarious Labrador fishery:

That your petitioner about Six years ago dealt with Messrs Thorne & Co. to the amount of £300 and carried on the fishery on Labradore.

That the fishery was very bad that Season and your petr. fell back on his account upwards of £43.

That your petr. dealt the following year with the said Thorne & Co. but could not reduce the balance of the former year although giving him every fish petitioner caught.

That your petr. was refused supplies for his family and was therefore obliged to dispose of what little property he had for which he could not get but very little for.

That your petr. about three years ago dealt with Mr Wells at Labrador and that year the fishery almost totally failed and your petitioner did not catch sufficient fish to pay the wages – but petr. has since paid him some fish and owe him upwards of twenty-eight pounds.

That your petr. has a large helpless family who have no person to trust to but ptrs. labour.[32]

Shannahan pleaded to be declared insolvent so that he would not have to face prison.

Even relying on family labour could not guarantee success to planters in the Labrador fishery. For example, John Day, a Carbonear planter facing imprisonment for debt in 1848, explained to the Northern Circuit Court that the proceeds of his family's fishing trips to Labrador rarely covered the voyage's costs, and that his high credit and transportation costs left him vulnerable to falling fish prices. Low prices in 1848 'left Petitioner penniless and his family without fuel and without many of the commonest necessaries for the winter.'[33]

Occasional records of insolvency by planters who were not involved in the Labrador fishery indicate that it was risky for them to employ servants in any large-scale fishery. Six of the thirteen cases of insolvency which surfaced in the sample of writs from the Northern Circuit Court could not be identified with the Labrador fishery (see appendix B). Along with one list of British and Newfoundland creditors to an unidentified insolvent estate owing £2,784 (probably a merchant),[34] only two of these insolvencies indicate large-scale operations. In 1827, after he paid his crew their wages, William Mosdell still owed merchant Charles Cozens of Brigus £995 for current supplies, £1,400 on previous balances due to Cozens, and £61 in other debts. Against this total debt of £2,456 Mosdell could only balance assets of £700, including a schooner valued at £300, a fishing room and craft at £160, and outstanding debts of £100 owed to Mosdell.[35] The assets which came to light after he became insolvent suggest that John Way operated a large fishery. Way failed in 1848 when he could not pay his supplying merchants, Ridley, Harrison & Co., the £300 they demanded (see table 2).[36] Other insolvency cases of planters and fishermen which could not be associated with the Labrador fishery indicate that these were usually smaller operations. William Marshall's 1833 debts of £117 (including £43 to Thomas Foley, his current supplier), for example, far outweighed his £7 worth of fishing equipment.[37]

The disenchantment planters could experience as a result of expanding the scale of their capital investment in the fishery emerges in 'a Natives' 1846 parable entitled 'John, of "The Harp," or, "The Way

TABLE 1
Insolvent estate of John Meaney, 1843

Debts owed by Meaney		Assets of Meaney	
Gosse, Pack & Fryer	£1023.14.02	1. Debts owed to Meaney:	
McBride & Kerr	164.13.06	Edward Barrett	£ 5.16.00
Wm. Bemmister & Co.	5.00.00	Thomas Oats	5.14.00
George Forward	4.00.00	Henry Thistle	0.13.09
John Rourke	4.09.10	Moses King	1.11.06
Wm. Brown	6.00.00	Patrick Redmond	2.08.04
Edward Walmsley	8.00.00	Edward Doyle	1.03.00
James Skehan	5.00.00	John Cornish	0.16.16
James Wall	16.00.00	Patrick Cashman	2.06.06
J. & F. McCarthy	4.00.00	James Butler	1.04.08
J. Peters	4.10.00	Richard Dunn	2.09.01
Stephen Brine	3.10.00	Edward Dunn	3.17.11
Punton & Munn	30.00.00	Robert Dunn	1.03.09
Nicholas Marshall	20.00.00	Jeremiah Dunn	1.09.11
		John Harris	2.19.08
		Walter Joyce	0.16.09
		Patrick Rourke	2.06.09
		George Butt	1.02.09
		Richard Doherty	0.18.03
		John Morea	1.01.09
		Thomas Fling	1.00.06
		Dennis Dunn	1.00.06
		Michael Wallace	1.04.00
		James Doyle	5.13.04
		Subtotal	48.18.00[a]
		2. Property	
		Fishing room &	
		premises at lab.	5.00.00
		6 puncheons	1.10.00
		2 skiffs	4.00.00
		1 cod Seine	5.00.00
		1 skiff	1.00.00
		Furniture and fishing	
		gear	
		under attachment	19.00.00
Total	£1298.17.06		£84.08.00

Source: PANL, GN5/B/19, box 74, file 4, writ no. 19
[a]Mistake in addition is in the original.

TABLE 2
Insolvent estate of John Way, 1848

Debts owed by Way		Assets of Way	
Ridley, Harrison		1/2 schooner *Success*	£125.00.0
& Co.	£322.17.1	Hire 1/2 schooner	
Francis Sheppard	5.00.0	last Spring	20.00.0
Robert Parsons	5.00.0	1 cod seine 11.00.0	
Samuel Bennett	2.00.0	1 caplin seine	5.00.0
Wm. Stirling, MD	1.11.6	1 lance bunt	4.10.0
Jonathan Parsons	1.05.0	3 fishing boats	13.00.0
Charles Parsons	0.05.0	1 stage lamp	0.01.0
Mrs Dixon	1.00.0	cod seine skiff	1.00.0
Thomas Dunford	0.10.0	2 skiffs rhodes and	
		1 old rope	1.00.0
Jonathan Kennedy	0.05.0	1 second hand rhode	
		and 8 fishing leads	1.12.0
		1 seine line, 6 jiggers	0.13.0
		3 grapnels, 3 creepers	
		1 howser	1.10.0
		1 mooring anchor, 2	
		dip net irons	0.13.6
		pews, gaffs, old hhd.	0.11.6
		6 hhd. salt	3.06.0
		1 boats compass	0.05.0
		boats kettles, 3	
		cod bags	1.11.0
		Interest in Harbour	
		Grace dwelling	30.00.0
		land + table	2.02.6
Total	£339.13.7		£222.15.6

Source: PANL, GN5/3/B/19, box 21, file 6, writ no. 38

to Get Dished."' In this account, merchant capital actually emerges as the venture capital of expanded scale of production, and increased employment of wage labour in the fishery. The story suggests that fifty years earlier, planters had had the right idea when they expanded the scale of their family operations to include spring sealing, that problems only began when merchants began to encourage planters to finance the building of large-decked schooners which required much labour and led to a heavier reliance on merchant credit. Before taking this new step, planters were simply hardy fishermen with good wives who provided for most of the household's needs from their own produce from the sea and garden.

John, a 'Native''s ideal, owned a small boat, catching and splitting his own fish, giving his fish and oil to his supplying merchant, and saving perhaps £150 over the years. From the labour of his own hands, John built his own house and garden, and raised some livestock. However, John's household's self-sufficiency disintegrated when he tried to expand his family's operation. John's merchant, 'Messrs Pale Seal & Co.,' encouraged him to set his son Tim up in a get-rich-quick scheme: he would borrow money to buy a schooner, the *Harp*, and make a lot of money from an expanded sealing operation.

A 'Native' found fault, not with the planter's desire to enlarge the scale of his operations, but rather with the merchant's subversion of the planter's household-oriented production and consumption. John could not earn enough from the capricious catches of the seal hunt, and he turned to the Labrador fishery to keep his capital employed during the summer. The proceeds from voyages there could not cover wages and schooner costs, and John ended his career impoverished and in debt.[38]

The parable of John the planter clearly idealized household petty production through reliance on family labour as the only way for planters to thrive on the northeast coast. Under other conditions the planters of Conception Bay either failed and joined the ranks of their fellow household producers, or they left the colony altogether. The recommendation of 'Delta,' another correspondent, was that remaining planters should not hire servants but should rely instead on their families' labour in both the inshore and Labrador fisheries.[39]

The fishery of the northeast coast of Newfoundland in the first half of the nineteenth century rested primarily on the labour of families within households, supplemented by servants at times when the family could not supply enough. The offspring of these households sought work as servants in the seal fishery and on the Labrador coast to buttress their families' income. Work outside the home was also a transitional stage for many, on the way to establishing their own households. Planters usually paid shares to the occasional servants they did hire, but the labour of the family proved to be the crucial underpinning of an economy based on household production.

Although some planters eventually became petty traders and shippers in the Labrador fishery, most remained resident fishermen who owned their own boats, equipment, and fishing rooms, and relied on family labour and merchant credit just as other fishermen did. Differentiation among planters proved to have very little potential in sparking a transition from a society dominated by family labour in household production to one based on the capitalist employment of labour – except during the boom times created by the Napoleonic Wars. The growth of a class of industrial-capitalist producers who might challenge merchant-capitalist hegemony in northeast-coast society ended with the wars. Differentiation among planters declined with the deepening post-1815 depression in fish markets, and the loss of the north shore fishery.

3

Household Agriculture

Fit for a fishery and little else – that is how Methodist missionaries felt about the northeast coast. People, in their opinion, depended almost solely on the fishery, not because merchants opposed agriculture but because, put bluntly, to 'the Agriculturalist Newfoundland promises nothing.' The fishing people of the northeast coast worked hard at growing vegetables and livestock, only to be frustrated by obstacles that were mind-numbing to contemplate. Frost and snow often arrived in October and did not depart until late May, sometimes June. Arctic ice blocked coastal waters and further lowered air temperatures. The interior lands were either bogs or lichen-covered, boulder-strewn barrens. People spent most of the year trudging through slush, which usually covered a shallow soil. Summer, as often as not, brought cold rain rather than warm sunshine. Agriculture was not totally impossible under these conditions, but required 'Herculean labour & pains' for the scantiest rewards of a few potatoes or turnips, and maybe a cow, sheep, or pig. The fishery, by comparison, was easy money.[1]

Despite the dismay northeast-coast settlers may have felt when confronted by such bleak land and weather, they set their hands to the fields. Fishing people knew they depended on merchant credit, that such reliance often ended in debt, and that it was important to find some way to control the amount of goods and services they had to take from the merchants. One way was to avoid hiring wage labour in the fishery. But residents turned to the soil as well for as much of their provisions as possible, in an effort to minimize their obligations to merchants. While some merchants and officials opposed settlement, producers in Newfoundland did attempt to diversify their economic activity through agriculture. Throughout the late eighteenth and early nineteenth centuries, Newfoundland fishing families explored ways to minimize the amount of provisions they had to secure on credit from merchants, including home production of consumer goods and food-stuffs. Indeed, the inability of merchants to provide fishing people with fresh garden produce forced settlers to try to provide these foodstuffs for themselves. Fishing families, merchants, and local government of-ficials all quickly discovered that the northeast coast's soil and climate allowed agriculture to serve only as a meagre supplement to the fishery.

This is not to say that West Country merchants did not worry in-itially that agricultural activity in Newfoundland might partially un-dercut their profits from the fishery's supply trade; some did. But as merchants restricted credit during the post-1815 recession, they looked to subsistence agriculture as a way in which families could provide themselves with foodstuffs to replace those no longer available on credit. Worried about the costs of credit, and faced with increased competition from the Americans and French in the European cod mar-kets, merchants hoped that a family based combination of fishing and cultivation would facilitate the production of salt fish at a cost that would undersell the American and French products. There had in fact always been a symbiotic relationship between agriculture and the fish-ery in Newfoundland. From the time of the proprietary colonies, New-foundland settlers farmed as best they could, and there is no evidence that merchants or government took any meaningful steps to stop them. While a far-removed colonial authority frowned on it, fish merchants came to accept farming as an indispensable support for the resident fishery.[2]

The merchant strategy with regard to agriculture at Newfoundland had been clearly stated in the 'Remarks of a Merchant on the New-foundland Fishery,' in an anonymous 1781 pamphlet. Its author told the Board of Trade that he 'would never Recommend any further En-couragement for Cultivation than the Inhabitants & Traders there may occasionally do for their own immediate purposes.' Merchants, he said, opposed diverting labour from the fishery and into any foolhardy at-tempt to develop large-scale cultivation in such a hostile environment.[3] The cooperation between merchant and settler in agricultural devel-opment, however, was neither equal nor non-exploitative. Merchants in the Newfoundland fishery were out to earn money from trade with fishing people. They accepted subsistence agriculture because it aided the creation of profit, not because it benefited settlers. By 1784 West

Country merchants would occasionally, and unsuccessfully, encourage the British government to prohibit the entry of American provisions into Newfoundland; they could accept people raising what local provisions they could, but they would not accept losing trade to American sources of supplies.[4]

While the state might agree as to the limited nature of agriculture in Newfoundland, it disagreed with the merchants on the question of American provisions. In 1785 Governor Campbell decided he must authorize public relief because people had not earned enough in the preceding fishing season to purchase provisions for themselves during the winter. While Campbell acknowledged that he did this to protect merchants' premises from attack by hungry people, he criticized merchants who continued to oppose the entry of American provisions. When merchants of the English West Country port of Poole mounted a petition against American imports, the governor's commentary suggested that the merchants' unwillingness to compete against cheaper American provisions represented nothing more than a blatant attempt to impose monopoly in the pricing of provisions on a distressed fishing populace.[5]

Fishing people needed to eat, and they depended on imported provisions to enable them to do so. Yet merchants saw the sale of such foodstuffs as a lucrative part of their business, and they were not willing to see settlers have access to competitive alternate sources. The British Board of Trade and the local governors allowed American supplies to enter the Newfoundland market, but this did not stop merchants from restricting the supply of food to families whose earnings in the fishery allowed them some hope of repaying. A season of poor catches or prices did not obviate the demands of empty stomachs, but that mattered little to merchants who worried more about carrying debt than feeding clients. Thus, a series of bad years in the fishery might see some clients denied credit for food by their merchants, and more successful planters might find that they had to pay higher prices to cover merchants' losses to their less fortunate neighbours. Hunger or debt were powerful motivators which encouraged fishing people to take up the rake and hoe. Those who could not raise enough to live on turned to the governors for relief.[6]

This situation led British officials in Newfoundland to think more positively about settlers' agricultural rights. British legislation forbade the engrossment of land for agriculture and the erection of structures for other than fishing purposes. Thomas Skerrett, brigadier-general of the garrison forces, suggested that the British government allow fishermen to enclose land 'provided it does not interfere with the Fishing grounds, and it is extended only, to the feeding of a Cow, or a Pig, and the planting of a few potatoes.' He thought such laws preventing enclosure had made sense when the British government hoped to preserve a ship fishery by compelling fishermen to return to the British Isles each year, but now government could not ignore the subsistence needs of residents. Skerrett emphasized that fishermen must be encouraged to raise potatoes in order to avoid the yearly wintertime distress and threat of famine.[7]

By 1803 both merchants and government acted to deal with the problem of provisions and agriculture. The merchants of Poole, Dartmouth, Teignmouth, and Bristol decided that money could be made carrying American provisions to Newfoundland – they petitioned for and received permission from the Board of Trade to import salt meats into the island. At the same time, Governor Gambier decided to allow leases of land for the purposes of cultivating gardens. This system, reported Governor Gower in 1804, applied mainly to the immediate neighbourhood of St John's because it was only there that government had usually enforced prohibitions against enclosure. Gower felt that the natural limits of agriculture in Newfoundland would confine it to a role complementary to the fishery.[8] He therefore proposed that the British government authorize year-round imports of American provisions; allowing them only during the fishing season did not give merchants enough time to meet the resident population's requirements, and people thus faced the yearly prospect of winter famine. Local agriculture alone could not meet the needs of people whose main resource and occupation was the fishery.[9]

Gower continued to push for more vigorous government support of agriculture because he could see no other way for fishing families to find relief from high priced, scarce provisions. The importance of such measures was clear: subsistence agriculture subsidized merchant profit in the fish trade. As Gower explained, 'if the proposed measure should afford them the means of obtaining a cheaper subsistence than at present, it would enable them proportionately to render their produce on easier terms to the Merchant which would encourage more of that class to engage in the exportation of it, and extend its consumption to the rival fishery of New England.'[10] There was no possibility of agriculture injuring the fishery, he noted, since settlers could barely provide themselves with garden vegetables or even raise enough

[Handwritten margin notes:]

His telling of the debate it obscured by fact he does not distinguish discourses

[the fact he does not distinguish discourses

Discourse have very odd – surely much of this agric would have been worried about. Aren't the 'promoters' telling about a different kind of agric – enclosure etc?

fodder to feed government officials' horses. Gower was determined to break down any British government resistance to encouragement of the resident fishery. He became an early advocate of constitutional revision that would do away with anti-property rights legislation, and he particularly objected to the prohibition against enclosure of land for cultivation, arguing that he and previous governors allowed fishing settlers agricultural rights as a means of ensuring their survival and the prosperity of the fishery.[11]

In gaining government recognition of fishing people's cultivation efforts, Gower tried to avoid the arbitrariness often shown by government officials in dealing with the issue; but a later governor, John Thomas Duckworth, at first objected to any measure which might encourage a resident fishery. Duckworth wanted to see the revival of the migratory fishery, believing it to have been a nursery, or important training ground, for British seamen. Despite his early hostility, Duckworth soon learned that Newfoundland's fishery rested on subsistence agriculture. In 1811, for example, he had to cope with the persistent problem of residents' not being able to find enough imported provisions to survive a winter. Like his predecessors Duckworth came to accept that the resident fishery had become dominant and that resident fishing families could only survive with the support of subsistence agriculture, but that Newfoundland's soil and climate would not allow agriculture to interfere with the fish trade.[12]

Others agreed. Methodist missionary Edmund Violet, for example, argued that it would do no harm if government allowed Newfoundland fishing families to cultivate the land because poor soil and climate set natural limits to the extent to which agriculture could compete with the fishery. Even capital investment would not improve growing conditions, Violet argued. The merchants had evaluated the potential viability of agricultural activity, but concluded that they could make money with far greater ease at their 'regular business, without cultivating rocks, or covering stones with earth,' and fishermen realized they could make more money trading fish than was possible trading potatoes.[13]

Despite these and other similar observations, the imperial government proved slow to change its views on Newfoundland agriculture. In 1812 the British government again tried to address a provisions shortage by allowing merchants in the Newfoundland trade to import American provisions into the island. Advocates of cultivation in Newfoundland argued that, rather than see the United States benefit from

trade to Newfoundland, the British government should grant fishermen full property rights as an experiment in encouraging them to
raise more food locally.[14]

Fishing families knew that the real problem was not competition
between agriculture and the fishery, but the fact that agriculture might
not even provide subsistence. Newfoundland's settlers had been cultivating the soil in defiance of British law for a long time, and they
still had not been able to do without American provisions. In June
of 1813 Governor Keats informed the Colonial Office that Newfoundlanders had again experienced a winter of near famine, and he observed
that year-round access to American provisions was a necessity because
merchants could not hope to import enough provisions at low prices
from either the British Isles or British North America. In many areas,
he reported, residents were forced to eat their seed potatoes when
their flour had run out. Even if cultivation received immediate official
encouragement, such action would be too late for the approaching
winter.

In mid-summer of 1813 communities from around the island had
already run out of supplies, but plentiful imports from Great Britain
and Ireland in the fall averted the famine Keats had feared. The governor began to make grants of land in the St John's area to help ease
the provisions shortage. He did not grant leases in the outports because
the surrogates there paid little attention to past regulations against
cultivation. Consistent with previous governors' acceptance of agriculture as a necessary subsidy to the fishery, Keats limited grants to
four acres so that every fishing family might raise its own potatoes,
vegetables, hay, and oats. In the outports he continued to observe the
local policy of allowing fishing families to squat on Crown lands so
that they could raise garden vegetables.[15]

The post-1815 depression in the fishery further encouraged local
agriculture. Depressed markets for fish caused by the increased production of the French and the Americans led merchants to restrict
credit to planters, leaving people with scant means with which to pay
for their winter's supply. By 1817 famine had again become a serious
prospect.[16] Post-war depression raised a new spectre before the eyes
of British officials: during the winter of 1816–17 threats of popular
violence forced the senior naval officer on the Newfoundland station,
Captain David Buchan, to issue provisions to local fishermen. Governor Pickmore, then in London, agreed that relief should be provided
as a temporary expedient, but indicated to British authorities that New-

foundland's surplus population would have to be removed because the economy was not likely to improve soon.[17]

The situation was dangerous. Government relief measures had not stopped people from threatening mercantile premises at Carbonear and Harbour Grace as they searched for food, and Pickmore was sitting on a powder-keg of discontent. British authorities did not want to authorize funds for relief, and their refusal left people without food or the immediate means of escape from the island. Despite the colonial secretary's admonishment to relieve people through public works, Pickmore had to issue stores. The problem he faced was that, without the fishery there was no useful work in which the governor could employ people. Cultivation and cutting wood might subsidize the fishery, but they were not areas in which residents could find full-time subsistence. Pickmore resorted to sending paupers out of the island on ships bound for colonial and British ports.[18]

While the merchants of Poole offered to supply cheap American provisions of bread, flour, Indian corn, and livestock for one season only, officials within the Board of Trade began to consider whether or not the Newfoundland governors had been correct all along in hoping that expanded subsistence agriculture might solve the provisions problem. To be sure, such encouragement would depart from past Board of Trade policy, but they now finally admitted that a resident fishery was a *fait accompli* in Newfoundland, and that agriculture might perhaps meet its needs. The board suggested that Newfoundlanders turn to animal husbandry. It also decided to allow the governors to lease additional small lots of land, cautioning that such land was to be used only by fishing families for their own support. Merchants were not to be allowed to engross large amounts of land which might in any way prevent fishing people's free access to it.[19] This change of heart was a response to significant public pressure. In Conception Bay crowds gathered in January 1817 to seize provisions from merchants.[20] Deprived of credit and supplies by merchants wary of the state of the market for salt cod, and unable to obtain enough to eat locally, people took what they needed for themselves. At both Carbonear and Harbour Grace, fishing servants broke into merchant stores while military and civil authorities stood by, unwilling to prevent them.[21]

A detailed examination of the 'riot' of 1816–17 shows that the fishery's labour force actively shaped government relief and agriculture policy. The crisis began on 6 November 1816 when Matthew Stevenson, clerk of the court at Harbour Grace, issued an order to merchants and planters. They were to withhold, from any servant who did not qualify for credit for winter provisions, £4 from any wages due him from his employment following the fishing season. This money was to be used to pay passage for such servants to America or Great Britain. The plan failed and as a result Conception Bay ended up with large numbers of unemployed servants to whom merchants were unwilling to extend winter supplies. In January 1817 the Court of Session ordered a meeting of Harbour Grace's planters and merchants which commanded that the most distressed servants report for shipment to St John's, and from there out of the island; failure to do so would bring a flogging and gaol with only bread and water for the winter.[22]

Refusing to be treated like convicts, servants ignored the order and sustained their decision by collectively seizing food for themselves early in 1817. This action alarmed Conception Bay authorities who could take no action until June when the ice broke, and so allowed a summons for help to be sent to St John's. The magistrates then sent a message that seventy to eighty men had been roaming the bay since 3 February armed with guns and sticks, seizing whatever food they could find. The magistrates had allowed the crowd to take the food to avoid bloodshed. Now they wanted aid to stop the plunder.[23]

The merchants' initial refusal to give the servants provisions on credit provoked the mob. The storekeeper of Patten, Graham & Co., Duncan McKellar, stated that when the mob approached him at the store at Bareneed on 3 February, one of its leaders, 'Nicholas Nevil shaking his hand in my face, said that as we had not given him provisions at the fall of the year, insinuated that he would have it by force.' On hearing that McKellar hoarded food in his house, the mob deputed five to six men to search it, and they found in the bedroom bread and pork, which they took.[24]

The 'mob' was actually an organized response by servants to the provisions crisis. Surgeon Richard Shea reported that the servants selected a spokesperson, Thomas Cooney, to talk with him. Cooney stated that the people needed food, and Shea promised to do what he could to help. Other servants were not so orderly; one of their number, Thomas Walsh, thought the 'mob' ought to return to Port de Grave 'like Men' to search every residence for hoarded food.[25]

On 27 March merchant George Best reported that fishing servants, who had been organizing in Conception Bay for some time, twice searched his house for food. Two deputies from the mob, Walsh and

Ryan, approached Best first, warning him that they heard he was hoarding food, and demanding that he permit a search. Best let them in when the mob threatened to break his door, but they found nothing. They then left to get a barrel of potatoes from Best's partner, merchant Charles Cozens.[26]

Fishing servants who searched for food during the winter of 1816–17 observed a form of collective self-discipline which belied any notion that they were simply a disorganized rabble. This discipline is well illustrated by the 'mob's' quest for food at Charles Cozens's premises. Cozens tried to explain that he only had a barrel of flour for the use of his own family. Two fishing servants, Thomas Trehea and John Murphy, believed him. Walsh, however, argued that because the mob had received something from every other place, Cozens should contribute as well. The next day, after visiting Best, the mob returned demanding potatoes. Cozens said he had none, but the crowd forced his store and took potatoes and a barrel of pork. When Walsh tried to incite the crowd to take as much as they could, Trehea reminded him that servants wanted only their fair share of food. Walsh fled Trehea's authority.[27]

[margin note: Pressing all buttons]

English merchants would not send their ships out to Newfoundland with goods for the next fishing season until government could guarantee their property's safety against these mobs of fishing servants. Bristol merchants argued that the British government would have to find some permanent remedy to the provisions crisis because the postwar depression in the fish trade was likely to continue. Like their Bristol counterparts, Poole merchants in the Newfoundland trade made it clear that under no circumstances would they ship provisions to anyone in Newfoundland until the government guaranteed the security of their premises. They not only wanted the government to remove unemployed servants to other colonies, they also favoured wider encouragement by government for subsistence agriculture.

Poole merchants opposed only colonization schemes which they felt the northeast coast's landward resources could not sustain.[28] English merchants did not object to agriculture, but they could not see how anyone could pay back credit through agricultural pursuits only. If the Newfoundland economy was to recover from the depression, they believed, recovery would have to rest on fishing families' labour, supported and subsidized by nonmarket agricultural activity. Such a policy lay behind the Colonial Office's decision to allow small leases.[29]

This decision reflected the governors' officially unsanctioned policy of allowing people to cultivate small parcels of land, a policy encouraged by the winter rioting. Conception Bay magistrates tried to arrest members of the mob, but only a few could be found.[30] Yet it was impossible to remove all the fishing servants forcibly, or to expect that merchants would donate food, as they had in Harbour Grace and Carbonear, for the purpose of 'quieting the minds of the people.'[31] The surrogate at Harbour Grace responded to popular complaints about food shortages by stating that the governor was more determined than ever to let any men who wanted it 'enjoy what land they had enclosed and till'd for the use of raising vegetables for their families.'[32]

The investigation of a House of Commons select committee in 1817 added weight to the conviction of imperial officials that they had to recognize and encourage settler agriculture in Newfoundland. Poole merchant George Kemp Jr, who had worked for his family's business in Conception Bay, only cautioned against British sponsorship of full-scale agricultural colonization. People could live by combining fishing and farming, he said, but not by farming alone where soil could not be significantly improved even by manuring it with fish offal and seaweed. St John's merchant J.H. Attwood was of like mind, but he emphasized forcefully that merchants depended on fishing families raising most of their own provisions instead of relying on imports purchased on credit.[33]

The British government in 1817 had to accept two things as a result of the 'mob' action of 1816–17. First, a permanent population had established itself in Newfoundland. This population could not survive on earnings from the fishery alone; people had to make their living from a combination of cultivation and fishing. Second, unless government encouraged this latter objective, it would have to pay for the relief or the resettlement of the people. The British government had no desire to spend money on these options, or even on bounties to support the fish trade – merchants could not be expected to supply provisions on credit if there was no hope for a return on that credit. As a result the government decided, on the advice of the merchants, to encourage family based production subsidized by agriculture. British authorities hoped to minimize relief expenditures and to encourage fish producers to provide for their own consumption as much as possible, so that the fish trade would survive the post-Napoleonic Wars' restriction of credit by merchants.

Merchants had come to depend on farming by settlers to support the resident fishery. While imperial authorities had devised no means

for the systematic alienation of land in Newfoundland, both they and
the local officials were prepared to recognize the rights of residents
to the land they had squatted on. Government hesitated to establish
the legality of real property fully, but this had little to do with the
opposition to settler agriculture. Rather, authorities wanted to ensure
that valuable waterfront property in St John's (formerly reserved for
migratory fishermen to cure fish) was properly leased to raise revenue
for the island. Merchants had been encroaching on these rooms to
extend their stores and warehouses. The British parliament, however,
had passed legislation in 1811 which provided for the leasing of that
property to the satisfaction of most merchants. The acrimonious public
disputes between merchants and governors which had led to that legislation, unfortunately created the impression that government opposed property rights for Newfoundland residents.[34]

While fish merchants were content with their leases, small traders
(who could not afford the prices set by a bidding process) and professionals (excluded because only those directly involved in the fishery
could hold waterfront property) did not benefit from the legislation.
One of the professionals, Dr William Carson, already frustrated with
local authorities because he had been denied patronage, seized on the
property-rights issue as a means to strike back at government. He
was further aggrieved by the governors' opposition to alienating farm
land with anything more than small leases. Carson was all in favour
of agricultural improvement, but wanted it to proceed in ways that
would benefit some more than others. While the governors could not
see how Newfoundland's meagre resources could support a landed
gentry, becoming part of such a landed gentry was exactly what the
doctor wanted. Fishing families should not have rights to their own
landed property, he maintained, but rather should be forced to work
on estates held by people such as himself. Governor Keats disappointed
Carson in 1813 when he refused to grant him a large tract of rent-
free land in the St John's area.[35]

Carson was a loud-mouth, singing clearly in the cacophony of demands for local taxation and constitutional reform that was growing
after 1818. With petty trader Patrick Morris St John's, he called for
improvement in Newfoundland through systematic agricultural colonization, and he tried to build popular support by crusading against
supposed government opposition to the property rights of the 'people.'[36] The governor, Sir Charles Hamilton, opposed their demands,
not because he did not support gentry landholding but because he

knew it could not work in Newfoundland. The governor constantly faced the problem of finding relief for people who had been cultivating the soil for decades to little avail. He could see no utility in putting too much faith in agriculture if Newfoundland's soil and climate could not even support regular potato crops. Agricultural schemes would only draw labour from the fishery (the only means of paying for imported goods), encourage further colonization, and result in more mouths to feed.[37]

Carson and Morris argued, however, that if people could survive through a combination of farming and fishing, one only had to imagine the prosperity to be had by bringing Newfoundland's millions of unused acres into production. The labouring classes of Newfoundland could find their provisions locally; they could escape the yoke of expensive imports. Hamilton saw through the artifice of the reformers. There was simply not enough hope of agricultural surpluses to allow a gentry to thrive on the backs of the potential tenantry already established in the coastal fisheries, let alone a tenantry on landed estates in the island's interior. Far better to give fishing families small plots on which they could build a hut and plant a garden.[38]

Nevertheless, imperial officials became interested in Carson's and Morris's rhetoric. These British administrators were growing weary of the constant provisions crisis in the fishery. The British government now fully believed that agriculture could solve Newfoundland's problems. As the French and Americans captured more of the markets for fish, the British felt that Newfoundland must respond with a cheaper product which would undercut the competition. Yet the imperial authorities did not want to subsidize the industry, recognizing that fishing people could not live on the restricted credit merchants provided in the context of depressed markets. Imperial officials came to believe that the Newfoundland product could be made cheaper if labour costs were subsidized with cheap American provisions and local agricultural production, which would also serve to lessen the burden of relief on government. In the end, therefore, they recommended that all restrictions be lifted from agriculture so that fishing families might be more fully employed in year-round activities.[39]

The Newfoundland Judicature and Fisheries Acts of 1824 represented an official break with past imperial objections to property rights at Newfoundland. Together these acts established a judiciary independent of the governors' authority, removed any remaining restraints on the real property rights of Newfoundlanders both in agriculture and

Did this really matter given squatter practice?

the fishery, and empowered the governors to lease, sell, and dispose
of unused land.[40] In 1825 a new governor, Thomas Cochrane, arrived
to implement the new system. He took an active role in encouraging
agricultural pursuits by fishermen, and over the next five years, tried
to lessen the colony's economic problems by sponsoring public works
and encouraging people to raise crops and livestock for the local market.[41] Despite all such encouragement, farming remained generally
small-scale and was unable to provide even a limited sustenance for
many of its practitioners. Families' agriculture was not adequate to
meet the provisions shortages caused by credit restrictions, which in
turn were caused by the continuing depression in the fishery. Governor
Cochrane found that people continued to come to him for relief in
order to survive.[42]

Government was already learning that without its help fishing families' limited cultivation was not enough to sustain them when merchants refused sufficient credit for winter supplies. Officials at Newfoundland began to relieve fishing families only after decades of
encouraging (jointly with merchants) agricultural development in association with the fishery. If the merchants and the governments of
Newfoundland and London showed little enthusiasm for large-scale
agricultural settlement, this was only because every good assessment
of Newfoundland's soil and climate indicated that northeast-coast conditions would not sustain it.

Indeed, even if official regulation had attempted to prohibit agriculture, fishing servants had demonstrated, through their collective
actions in times of crisis, that they cared little for such laws. If merchants were going to tighten credit when depression struck the fish
trade, then servants were determined to take the food they needed.
However, this was only a temporary solution. Servants could not import food themselves, and ultimately they had to depend on fish merchants' credit over the long term. To ease the tension inherent in thess
conditions, government officials in both Newfoundland and London
accepted that the best compromise was to recognize what merchants
and resident fishing people had long realized: namely, that the fishery
at Newfoundland was best prosecuted through petty production supported by the subsistence cultivation of fishing families. Merchants
accepted agriculture because they also now knew what fishing families
had discovered earlier: that no amount of cultivation could force
enough surplus from Newfoundland's relatively barren soil and harsh
climate to allow an escape from dependence on merchant capital.

4

Women in Household Production

'Whore,' cried Mary Kough as she threw stones at Catharine Gould, threatening all the while to kill her.[1] Why would Kough so viciously attack Gould? The answer is a poignant indicator of the limited potential of fishing people's agriculture in 1840: Kough attacked Gould because the latter tried to prevent her from stealing topsoil from the garden of Gould's son-in-law at Carbonear. One woman was almost willing to murder another for the sake of a thin layer of manure and peatbog, a layer, however, which took an entire summer to prepare. Soil, crops and animals were scarce commodities in a society in which fishing families depended almost solely on fish to lessen their dependence on merchant credit, let alone stimulate a diversified domestic economy.

Kough's fight with Gould also suggests that women saw family agriculture as their special responsibility – by no means an unusual circumstance. In other parts of British North America patriarchal relationships between men and women, both in law and practice, differentiated household production. Women's principal duty was to care for the family's needs, not produce for the market. Such duty meant that women provided for their families with the produce from their gardens. The farming of Upper Canadian women did much more than free male labour for market production, however. When the women produced more than their families required, they exchanged produce from household to household, thus encouraging the growth of domestic industrial diversification, particularly in textiles, clothing, poultry, and dairy products.[2]

It is useful to contrast the success of women's agriculture in Upper Canada with the pathos of Kough and Gould fighting over topsoil in Newfoundland. First, although Newfoundland households shared the same legal structures of patriarchy with the rest of the Anglo-American world, the fishery's production and marketing requirements demanded a much closer integration of female and male labour. The northeast-coast economy was consequently much more restricted. Moreover, women's subsistence production found little encouragement in the coast's limited landward resources, factors that did not bode well for the development of the northeast-coast domestic economy. Second, the struggle for survival by fishing households partially eclipsed their formal patriarchal structure. Women's indispensable role in the production of cod and their domination of the households' subsistence challenged formal male authority. Yet such a challenge ultimately remained subordinate to the struggle for survival, and tensions between households ultimately reinforced the bonds between men and women within the families.

Women played a critical role in the Newfoundland fisheries' transition from a migratory to a resident industry. Settlement developed partially through marriages between migrant fishing servants and the young women of resident families who had stayed on after the failure of the proprietary colonies. Other women, who came to Newfoundland as the domestic servants of government officials, sometimes stayed to marry resident fishermen. As more planters engaged in the fishery, they also may have brought out female servants from Great Britain. These servants, like their male counterparts, often married into their masters' families.

Until the late eighteenth century the resident population grew slowly because of the imbalance favouring males over females in the seasonal migration of labour.[3] Women servants were more likely to stay on the island, marry, and produce daughters who would in turn marry the next generation of male servants.[4] Thus, the female population provides an index of permanence for the resident population: the merging of female to male sex ratios in the late eighteenth and early nineteenth centuries indicated the establishment of permanent fishing communities.[5] A process of transatlantic family migration in fact established the patriarchal family structure of West Country English society in Newfoundland in the early modern period.[6]

INHERITANCE

The increasing numbers of settlers at Newfoundland raised questions concerning which conventions would govern the transfer of settled property between generations. Such inheritance rules played a crucial role in defining women's place in society. Women who joined fishing households entered a patriarchal family structure defined both in law and practice by inheritance, as it was in other households of the Anglo-American world of this period.[7] Widows usually inherited little property from their deceased husbands' estates. Those who did, like Mary Sheppard in 1788, were not allowed to alienate such inheritances from the deceased husband's patriarchal line. During a dispute with her son Adam, Mary Sheppard produced in court a copy of her husband's will, which bequeathed 'what goods he may Died possessed of unto the Petitioner his Widow for her use during her life and after her Death to the said Adam and his heirs.'[8] Clearly Martin Sheppard did not intend to let his wife take property away from his heirs if she joined another household. In this case, Martin's property belonged to his son, not his wife. Jane Mardon also found out about this practice in 1789 after her husband, John LeCoux, a former Jersey man, died leaving her their fishing room at Western Bay. James, a son Jane did not know about, showed up from Jersey claiming the room, and the Surrogate Court awarded it to him, allowing Jane only one-third of any proceeds from the lease of the property.[9]

As it was elsewhere in the Anglo-American world, when married daughters inherited property from their fathers, their husbands assumed ownership. This emerged in a number of disputes over wills in the Supreme Court. In 1817 Nicholas Newell, a St John's planter, testified that he was married to Frances, the eldest daughter of Robert Mugford, a deceased Port de Grave planter. By Mugford's 1793 will, his wife, Ann, inherited his property for her natural life, but then it was to pass to Frances. Ann died in 1813, but left no property to Frances; she had sold the land to John Walsh. Chief Justice Tucker returned the property to the Newells. Similarly, in a dispute over the estate of Thomas Thistle at Harbour Grace, the Supreme Court carefully ensured that the use of their father's property was properly passed on to Thistle's four daughters, but through the ownership of the daughters' husbands.[10]

Beginning in the late eighteenth century, men usually governed the process of intergenerational household formation. Court records show that, throughout the late eighteenth and early nineteenth centuries, men in Newfoundland inherited property in lieu of their mothers or wives, usually with some provision that they care for their mother or mother-in-law. Occasionally, if widows inherited property, it was for their lifetime only and was not to be alienated from the family line of their deceased husbands.[11]

Widows who received such awards usually affiliated with the household of a son-in-law if their own sons were hostile to their estate claims. Jane Smith testified in 1827 that her sons would allow her nothing from her husband's estate, forcing her to live with an impoverished son-in-law.[12] Being dependent on the good will of a son could be disastrous for a widow. In 1838 Mary Murphy petitioned Assistant Supreme Court Judge E.B. Brenton, complaining about the treatment she had received for nearly ten years from her stepson, Dennis Murphy. Mary Murphy's husband, Patrick, died in 1827, leaving her with two children and a ninety-year-old mother to support. Dennis Murphy cruelly told Mary that his household would not support her, and that he would allow her to live under his roof only so long as she worked as he required and provided her own food. Eventually Dennis kicked Mary out of the Murphy house, and she moved to St John's to work as a servant in order to provide for her mother and children. The court awarded Murphy only a little over a pound from Patrick's estate.[13]

Some women do appear to have assumed important leadership roles within their households if there was no man to assume ownership of a household's inheritance. Widows left as custodians of their husbands' property could take on the important economic roles of estate manager and temporary household head, though they were legally restricted from selling the property.[14] Ann Brazill assumed direction of her son's interests when he died a planter's servant in 1787. The planter, Patrick White of Bell Island, paid Brazill, upon her suit, the wages he owed her deceased son, John Butler. In another case Mrs Thomas Thirts of Harbour Grace tried to protect her claim to a fishing room from the intrusion of a neighbouring planter family, Elizabeth Webber and her sons. The court allowed Webber – a widow left with eight children to care for – to build her stage on the property. She had to contend with her eldest son Charles's attempt to mortgage their property to merchant Thomas Lewis to pay his own debts, but the court allowed Charles only an eighth part of the total property, and gave Mrs Webber a third of her late husband's land as her sole property.[15]

Women without male relatives also asserted their inheritance rights to the fishing equipment of their deceased husbands. In 1822, for example, Eleanor Canty petitioned surrogate Toup Nicholas to have her husband's share of a caplin seine turned over to her after his death. Timothy Canty had bought the seine in partnership with Martin Casey at Harbour Grace. When Timothy died, Casey tried to steal out of the Harbour with the seine, only to have Eleanor demand her share from him. When Casey refused her request with much verbal abuse, Eleanor Canty had the Surrogate Court enforce her demand.[16]

Some women, like Ann Protch, appear to have earned their living by leasing such property to other fishermen rather than conducting a fishery themselves. Jane Cook, who acted for Protch in the matter, also handled their account with merchant James MacBraire, and rented other property at Harbour Grace to John Clements. Indeed, a number of cases appear in the Surrogate Court showing that women managed property and accounts left to them by their husbands.[17]

Throughout the first half of the nineteenth century women continued to pursue, through the courts, their rights as leaders within their own fishing households. Sometimes a woman submitted her case to the court by way of a petition presented by a son. In 1827 Ann Mugford petitioned the Northern Circuit Court through her son, Richard Fillier. She wanted the fishing room she inherited from her deceased husband, Richard, protected from the claims of Port de Grave merchant William Dunning. Dunning's supposed credit to Fillier was a thirteen-year-old claim of which Mugford knew nothing. Mugford made it clear to the court that she had no intention of giving up her fishing room (which she leased to other fishermen), especially considering that at the time her husband supposedly contracted the debt, Dunning had refused to grant them any more supplies, being dissatisfied with Fillier's return of fish for credit in the previous fishing season. Other women, like Ann Taylor in 1829, sued to protect their inheritance rights from brothers who tried to exclude their sisters from the division of their fathers' estates.[18]

THE FISHERY

Newfoundland fishing households were similar to other parts of Anglo-America in the formal structure of patriarchy within the family, but they were dissimilar – at least when compared with Upper Canada – in the greater degree to which Newfoundland households integrated women's work into market production. Patriarchy in Upper Canada reinforced a relatively strict division of labour between men and women. Only during the land-clearing phase of the initial settlement did women become extensively involved in the market-oriented work of preparing the land for crop production.[19] During the fishing season, however, all family members, both male and female, gave precedence to making salt fish. The household's women had to stand ready to cure fish when the men brought it ashore. Indeed, the rhythms of men's work in the catching of fish dictated the pattern of women's and children's labour during the fishing season. As long as there were fish to be caught and made, women's subsistence production remained secondary and subordinate to the household's principal productive activity: the supply of fish and oil for trade in truck with merchants. Consequently, whatever women were doing when the fish arrived to be cured had to be dropped immediately. Curing left little energy for other work.[20]

So important to the fishery was the curing work of women that newspaper correspondents sometimes reminded their readers that women in shore crews were as essential to the industry as fishermen.[21] The fishery required so much of women's labour that it also left them little time to produce surpluses from their subsistence activities, surpluses that might be exchanged locally as they were in Upper Canada. This demand for women's labour, rather than any change in the labour process itself, was the most significant change in the fishery from the seventeenth to early nineteenth centuries. Only the scale of activity changed as smaller family based operations overtook the sometimes larger ones operated by merchants who engaged directly in the fishery.

In first establishing their operations, fishing people had to build stages, or wharves, at which they could tie up boats and unload fish. A stage often included a shed in which to store the salt for preserving the fish and tables upon which workers processed the fish. Near the stage was a train vat, a container used to hold cod livers while the sun rendered them into oil. If there were no good beaches on which to spread the fish for drying, fishing people would also have to construct flakes (often large platforms of wood, bark, and boughs) for this purpose.

Once the shore facilities were finished, men would catch bait, then row their boats to the inshore fishing grounds. To catch the fish they used baited hooks on lines up to thirty fathoms long. If full of cod, these lines could weigh fifty to a hundred pounds, but usually weighed between five and ten. The lighter weights were still no small burden considering the repetitiveness of entire days spent fishing: constantly pulling lines in to unhook fish, re-baiting and re-lowering them, pausing only to row a fully laden boat to shore, then unloading and perhaps heading out again. Fishermen used prongs to throw the fish upon the stage, where the young people of the shore crew would lay the

fish on tables. A header would then slit the fish's belly, extracting
the liver for rendering and discarding the guts and as offal. Next, a
splitter would take the fish, cut it abroad, and discard the bones. A
salter would then layer the fish in piles, with salt, for curing. Finally,
the fish would be spread out to dry in the sun and air by the shore
crew, joined at the end of the season by the boat crews who helped
guard the fish against rain and burn from overexposure to the sun.[22]

The rise of the resident family fishery saw women and children take
on much of the work formerly performed by headers, splitters, and
salters, but otherwise the work of the fishery remained unchanged.
The needs of the fishery consumed both men and women. Men went
out in their bait boats on Monday mornings to catch caplin or squid
for the week's fishing. For the rest of the week they went to the
fishing grounds to catch fish, bringing in a loaded boat as soon as
possible, unloading it in exchange for more bait, then returning immediately to the fishing grounds. Most did not stop for sleep as long
as there were fish to catch. A Wesleyan missionary reported, 'I have
heard fishermen say they have not had their fishing boots off for a
week together.'[23]

Women, as their male relatives' shore crew, had to match the men's
frantic work pace. The shore crew unloaded the fish, split and salted
it, and spread the pickled fish on flakes for drying. Then they had
a short break in which they could attend their domestic work, but
they had to watch the fish constantly, turning it frequently to prevent
sunburn. Dried fish had to be taken up and then stacked skin up so
that moisture would not damage it while the salt cod awaited carrying
to the merchants' stores. This work often kept women busy until
midnight.

The pressure of this shore work must have been intense. Merchants
paid for fish by the quality of its cure. The determinaton of that quality
in turn rested not so much on the manner in which the men caught
the fish, but rather on the women's ability to cut and salt it, to watch
the weather, to decide if fish needed to be turned or covered, and
whether it was ready to be taken up from the flakes for storage or
marketing.

The other main labour of the people of the northeast coast, sealing,
was an exclusively male occupation. Preceding the inshore fishing season, the seal fishery began soon after 21 March. Schooners, from Conception Bay for the most part, travelled to the pack ice where the
seal herds could be found. If the sealers were lucky, seals would be

sighted only two to four miles from their ships. If not, the sealers would have to travel even farther to reach the seals by jumping from ice pan to ice pan. Men faced the constant hazard of missing pans or mistaking slush for solid ice, often losing their lives to the north Atlantic's frigid waters. While on the ice, sealers could also be set upon by fierce snowstorms or fogs, making return to their ships almost impossible. Work was a constant process of locating the seals, slaughtering them, and then moving on, all the while looking over one's shoulder in constant surveillance of weather and ice conditions, often trying to return to one's ship guided only by the sound of a gunshot or whistle. Killing seals was a baptism of blood as sealers crushed the animals' skulls with gaff poles and stripped the pelts and fat as blood spurted over the sealers and the surrounding ice.[24]

Women did not work in the seal hunt, but they could find employment in the shore crews of other households. On at least one occasion a fisherman contracted his wife's labour to cure another man's fish.[25] On occasion a man could bargain to hire both his and his wife's labour to a planter involved in the Labrador fishery. Patrick Loughlan hired himself out to Edward Guerney of Carbonear to work as a fish splitter in return for wages of £26 and four quintals of fish. Patrick's wife, Mary, was to accompany him to help make the fish for £8 wages.[26]

The great attraction of female labour was that it commanded much lower wages. In the course of an 1833 court action, Joseph Pippy gave a statement concerning his summer fishery at Labrador. Pippy had four sharemen as well as three male wage servants who received wages of £12, £20 and £21 respectively. Pippy gave his wife £8 for her work as a salter. Pippy's two other female servants – daughter Lydia Pippy and Ann Coke – received only £5 and £6 each. Pippy enjoyed the double advantages of cheap female labour that was also subject to the discipline of the family structure.[27]

Women employed in the Labrador fishery performed shore work for planters who did not have enough family labour to draw on. In 1835, for example, Mary Reed complained to the Northern Circuit Court about nonpayment of wages. Her testimony reveals that Reed agreed to work for Thomas Davis and his five sharemen on the Labrador coast. Reed cured and made all the fish they caught in exchange for £13 wages. When Davis refused to pay her, and the sheriff could not find enough property of Davis's to secure the wages, Reed charged that Harbour Grace fish merchant Charles Nuttall had taken the fish and oil, as Davis's current supplier, in a conspiracy to defraud her of her wages. The court ordered Nuttall to pay £16 in wages and damages to Reed.[28]

Although women's labour was an integral part of the staple production of fish, the formal patriarchal organization of the cod trade remained unchallenged. Society did not recognize women as partners in their households' dealings with merchants. Out of the hundreds of suits by merchants to recover outstanding account balances in the Surrogate Courts at Harbour Grace between 1785 and 1824, there exists only one in which a husband and wife were the defendants in such an action. In 1794 Richard Cornish sued Grace and John Holmes 'for recovering a Book Debt for Goods sold and Delivered in the year 1791 – amounting to Twenty-Eight pounds.' Both parties compromised, with the Holmeses agreeing to pay the debt off in six yearly payments on a mortgage of their fishing room at Adams Cove in Conception Bay.[29] In almost all the other cases, merchants sued single men or small partnerships of two or three men.

While official authority did not recognize women's partnerships with their male relatives in fishing households, women did appear in court to defend their households' interests. Nancy Daw of Port de Grave, in 1831, tried to protect her household from the effects of a judgment of about two pounds levied against her husband, George, in 1830. A sheriff's attachment saw part of the Daws fishing room sold to John Daw. Nancy Daw informed the court that they had plenty of other property to satisfy the debt, that selling part of their room resulted in a 'great and manifest injury ... in as much as her said Husband has not any place to dry and cure his fish by which he makes a living for his family.'[30]

While the record does not reveal the outcome of Nancy Daw's plea, the language of it says much about her place within the patriarchal structure of outport society. Her appearance in court suggests that Nancy Daw played an important role in the management of her fishing household's affairs. But her reference to her husband's not having a place to dry his fish indicates that Nancy Daw recognized her household's patriarchal legal status. Daw was part of a society in which a woman's labour was simply a unit of production in the household, and her labour, at least in the court's eyes, was seen as part of her husband's estate.

The lack of legal or mercantile recognition of the (unequal) partnership of women in the production of the cod staple does not mean that women did not force their presence to be recognized. This is dramatically illustrated in the case of James Lundrigan in 1818. Lundrigan was a poor planter who was harassed by his supplying merchant for falling behind on his account payments, and eventually whipped at the naval surrogate's orders in 1820 for not yielding his property to a writ of attachment.[31] The severity of the punishment inflicted on him was a result of a threat by Lundrigan's wife against two constables, Kelly and Moors. Moors testified that when he and Kelly went to the Lundrigan plantation at Port de Grave 'there was nobody in the house but a woman and one or two children. The woman who was Plaintiff's Wife, desired Kelly to be gone or she would blow his brains out.'[32]

The Lundrigan affair suggests the displeasure of the state concerning the manner in which the particular importance of women to the fishery had begun to challenge male authority in fishing households. Except for three whippings of male servants for theft and fraud in 1787–8, the surrogate courts had never used corporal punishment as a sentence for anyone in the fishery until Lundrigan. Captain Buchan, the surrogate who ordered the whipping, had been acting governor during the store break-ins of 1816–17. He had hoped that the government's encouragement of the family fishery would replace the disorder of unemployed servants with the quiescence of family discipline under the authority of male household heads. But Buchan found in Lundrigan's wife no passive submission to a court order which threatened the loss of her household's means of survival. The naval captain also had to contend with the wife of another planter, Philip Butler, who would not yield household property to a writ of attachment for debt. Butler received a lashing as well. The two planters were whipped, not because of any personal act of contempt, but because they had failed to discipline their wives to allow the court's orderly settlement of debt disputes. Buchan, when faced by cases involving acts of contempt by men alone, usually ordered fines only.[33]

Although the courts expected male planters to ensure that their families did not disrupt the efficient payment of debt, women continued to assert their presence forcefully. In 1827, for example, when constables Edward Janes and William Legrow attempted to serve a writ on John Earles at Broad Cove, Earles, along with his wife, Ann, and his sister-in-law, beat the two. Ann defended her family's property with a hatchet, threatening to 'cleave him down' if constable William LeGrow came near her. Constable Janes reported that 'previous to Deponents demanding entrance into the house where the Calf and lambs so attached ... were concealed by the said John Earles – the Deponent read the Kings Writ of Execution to the said Jn. Earles who replied he did not care for the King's Writ the said Jn. Earles Wife Ann Earles standing by – said – shit on the King's Writ after this she took up a Hatchet in her hand in defiance of this Deponent as aforesaid.'[34]

Similarly, in 1836 Chief Constable James Sharp left Harbour Grace to help constable Jonathan Martin seize the fish of Samuel Pike at Musquito for payment of a debt, and as a result of a suit against Pike by merchant Peter Rogerson, Ann Pike, Samuel's wife, asked Sharp 'What brought you here you long son of a bitch.' Ann Pike then told the constables that she would die before she allowed them to take the fish in question. When a boat arrived to carry the fish to Harbour Grace, Ann threatened to smash the skulls of the boat's· crew with a rock, and then began to beat the constables. Finally, Sharp struck Ann Pike down with a stick. Upon this, Edward Pike, 'son of Edward' (probably a relative) approached the scene, while Samuel looked on and declared that, 'they were all mean low spirited people to permit the Fish to be taken away. The Mother of the said Edward Pike and two of his sisters were also upon the spot menacing and threatening the officers calling them all sorts of bad names.'[35]

Fishermen also used violence in an attempt to assert their personal authority within the household. Mary Barry, wife of fisherman Michael Barry of Harbour Grace, found this out in 1828 when a number of fishermen living in her house accused her of being a 'whore.' Mary Barry brought these men to court, where one David Bansfield – a fisherman who had lived with the Barrys for four years and served Michael at the Labrador fishery – denied the charge and attested to Mary's good character. This was too late to save Mary from a beating at Michael's hands earlier in the week, in addition to the terror of being forced to listen to Michael's debate with himself as to whether he would shoot her or split her skull with a hatchet. The court arrested Michael on Mary Barry's complaint.[36]

Ann Noel's position as wife of Carbonear planter John Noel did not protect her from her brother-in-law, Charles Noel, in 1835. Charles lived in Ann's household, according to some family agreement, until his own house could be built. After Ann refused to acknowledge that Charles owned any part of her (and John's) home, Charles swore that he would either cut Ann's throat or burn down the house with her in it. Charles chased Ann and her servant 'into the stairs of said House and exposed his Nakedness and his backside slapping the same and told complainant to kiss the latter he called Complainant a strumpet and whore and other such approbrious [sic] names.'[37]

Such threats could not be taken lightly. Patience Hussey, wife of fisherman William Hussey at Port de Grave, complained that she greatly feared for her life as her husband repeatedly beat her and threatened her life. William's brother John and his daughter Ann, in addition to a neighbour, Frederick Kenny, all testified that William Hussey had an insane paranoia that his wife meant to do him some harm. Kenny stated that he had begun to think it usual to hear Patience cry murder at her house each night.[38]

Female children and servants also had to contend with male violence. In 1844 Charlotte Bradbury complained that her father, John McLean, 'gave me a cut over the right Eye and bruised me in several parts of my body whereby I was unable for some hours to walk to my own home.'[39] Sarah Dalton stated in another case that her uncle, Thomas Dalton, struck her with a stick simply because he did not like the way she looked at him.[40] At Musquito Sarah Neary received such a violent beating at the hands of her master, Arthur Thomey's son Henry, that she refused to return to his service. In 1853 Edward Shanahan beat his servant, Catharine Chitman, for accidentally tipping over a cart of caplin. Catharine's mother, who had hired her out to Shanahan for four pounds, came to take Catharine home when she heard of the beating. The court did not think the beating totally unjustified, and allowed Catharine only some clothes bought for her by Shanahan. She had to forfeit her wages.[41]

Female servants in the Labrador fishery faced yet another problem. Mary Ryan, hired by Patrick Meaney of Musquito in 1832, Johanna Connors, hired by Kennedy Thomey of Musquito in 1835, and Eliza Mills, hired by John Burke's brother William in 1847, are examples of women who became pregnant as a result of their relationships with their employers while at Labrador. All these women had to sue for support of their illegitimate children on their return home.[42]

AGRICULTURE

The family's basic struggle for survival on the northeast coast ensured an essential mutuality between men and women in households, despite the presence of male violence. This commonalty arose from the labour demands of the fishery, but also because of the severe limits of agriculture. The subsistence activities in which women engaged – particularly in household-oriented agriculture – operated under very severe resource constraints. John Walsh reported in 1819 that in the post-war recession families avoided buying as much as possible from merchants, subsisting on their own fish and potatoes, and mending their clothes to stretch them along. When they were not working in the catching and drying of fish, both men and women did what they could to supply their families' needs, the merging of nonmarket and market production thus partially crossing gender lines. Men cut wood for fuel, boatbuilding and repair, or making utensils. Women made garments. In winter families would migrate into the shelter of woods near the coast where fuel could be more easily obtained. Aside from this, there was little for women to do because of the limited nature of the northeast coast's agricultural resources.[43]

Although such activity was limited there is evidence that the management and labour of the household's farming was the special preserve of female family members, when they were free from the overriding demands of the fishery. In an early Surrogate Court case, for example, Mary Cole of Colliers in Conception Bay sued Stephen Hunt for the illegal seizure of her son's cattle. The court ordered the cattle returned. Cole appeared to earn her living by being 'shipped' for the summer as a servant of Jason Ellison, but the exact nature of her work is not specified in the court records.[44] Johanna Healey, in 1829, made it clear in the settlement of the effects and property of her late husband, John, that she should inherit the family's garden and meadow, contrary to the claim of her son-in-law, Philip Meaney. While her husband was away at the Labrador fishery in 1836, Catherine Callahan looked after their garden. When James Counsell's pig got into it, she yoked the pig and vigorously protected her garden by threatening Counsell's wife, Mary, with a beating if she tried to take the pig. The Northern Circuit Court ordered Callahan to keep the peace. Meanwhile, in 1845 Ann French took her neighbour, Thomas French, to court for allowing his dog to run loose and harass the goats kept by local families. Ann felt it her duty to protect local livestock from this menace.[45]

In an 1845 dispute at Harbour Grace, court evidence made it clear that women were the experts when it came to the family's livestock. On his return from the Labrador fishery, George Heater's wife, Sophia, had informed him that a neighbour's labourer had killed their female goat when it entered the neighbour's garden. Sophia had a number of her women friends testify as to the value of the goat. She stated that after she had milked the goat she had properly yoked it. The court fined the labourer, Thomas Pine, for his action.[46] One reason women used intimidation to protect their property from neighbours and court officials (as in the case of the Earles' defence of their cattle at Broad Cove) may have been that the former threatened their gardens and animals either by trying to seize them to pay debts incurred in the fishery, or by allowing the animals to roam through their gardens.

The constraints imposed on women's subsistence activities by Newfoundland's relatively poor agricultural resources can be seen in court cases such as the one mentioned earlier, in which two women actually fought for possession of topsoil. Mary Kough had led two other women in raking topsoil off Gould's son-in-law's garden. A similar incident led to a fight between two other women in 1844: 'Mrs Dogherty called Mrs Cotter a Bitch and was returned the same language of abuse – when Mrs Dogherty spit in Mrs Cotter's face *and struck her with a stick and made Mrs Cotter's nose bleed* [court's emphasis], which Mrs Cotter returned ... they each claimed some sods cut by Mrs Cotter in the woods – they both claim the land where the sods were cut.'[47]

The tension implicit in women's efforts to provide for their families with such meagre resources made their gardens a focus for much violent confrontation between them. For example, in 1839 Eleanor Sullivan at Harbour Grace charged Susan Russell, wife of fisherman Patrick Russell, with assault. In her own defence Susan Russell told the justices of the peace that she only tried to prevent Sullivan from stealing roots from her garden.[48] Such conflict cut across even family lines. In 1853 at Marshall's Folly in Conception Bay, Mary Slade testified that she was both relative and neighbour to Rebecca Slade and Ann Slade. In August, she said, she heard the two arguing outside her house, and went to see what was the matter. At her door Mary Slade saw 'Rebecca and Ann Slade standing just outside their own doors – they were disputing a fowl laying an egg in the garden, when Ann said that she removed her own eggs from her back house for fear Rebecca should take them – and then Rebecca said a Rogue can't trust a Rogue – and then Ann ran towards Rebecca and said don't think that I am afraid to strike you – and she put her clenched hand up against Rebecca's face, she did so three times – they were not blows but pokes in the face.'[49]

This kind of violence could unite people of both genders within families, against other families. Edward Noftell, a fisherman of Broad

Cove, complained to the justices of the peace in 1845 that his brother
Nathaniel let a horse into his potato garden, destroying some of his
crop. Edward threw a rock at the horse, causing Nathaniel to strike
him. The two brothers began to fight. Nathaniel's wife, Mary, seeing
the fight, called on her daughter to strike Edward in the face with
stones. Upon the entry of Nathaniel's womenfolk into the fray, Edward's wife, Louisa, rushed to his defence, only to be struck down
by Mary with a spade. A third brother, James, watching the fight,
called to his own wife to help Louisa into his house while he returned
to digging his potatoes. Occasionally women also served as peace-
makers: in 1834 Luke Micheton told J.P. Thomas Danson that James
Titford would have beat him to death in a fight over manure, except
that Titford's wife, Mary, pulled James away from him.[50]

The tensions between the Slade women, or the fights which broke
out among the Noftells, are not so much an indication that interfamilial
squabbling was common on the northeast coast as they are of the
manner in which family solidarity shaped by desperation outweighed
any other consideration in the lives of women in fishing households.
The Slade women fought each other to make sure their own immediate
families got their fair share of eggs. The Noftel brothers' wives and
daughters came to their defence in a violent dispute over the potential
destruction of the family's all-important potato garden. Other women
battered each other over soil. These strong actions suggest that the
imperative influence shaping men and women's relationships within
fishing households was the constant effort required to supply the
households' subsistence needs.

The intrinsic interdependence of men and women within northeast-
coast households is further suggested by the absence of a certain type
of evidence. Elsewhere in British North America the press of the late
eighteenth and early nineteenth centuries commonly printed men's notices stating that they would no longer accept responsibility for the
debts of runaway wives.[51] Running away allowed many women to
break out of the restrictions of patriarchal household structures.[52] Notices of their action in the press indicate that in other colonies conflict
between men and women within families could break households
apart. Freeman Durham of Beverley in Upper Canada, for example,
warned in 1829 that he would no longer pay the debts of his wife,
Rachel, because 'some unhappy differences have lately arisen between
[us, and] ... we have mutually agreed to live separate, and apart from
each other.'[53] No similar notices exist in the surviving newspapers
of Conception Bay for the first half of the nineteenth century.

The very serious threat of famine entrenched the loyalty of northeast-coast women to the patriarchal household. John McGoun, commissioned by the governor in 1831 to survey the northeast coast, found wherever he went in Conception, Trinity, and Bonavista Bays, women huddling together with their children while their husbands roamed the coast looking for food or work. In most cases they had so little food they could only hang on from one meal to the next, none of the family having much energy to do anything else but wait for the next meal. McGoun found one well-off family in Spaniard's Bay whose integration of market and subsistence activities he recommended to all: 'they had a family of *ten children* the eldest and greatest part of whom were daughters. These however were the personifications of industry – unusually warm as the morning was they were hoeing away in the potato field like Irishmen utterly regardless of fatigue from being used to exertion – they were the only "hands" that their father took to *Sea.'*[54]

The survival of families on the northeast coast depended in part on the merger of market and subsistence activities by women in staple production. As settlement grew in Upper Canada, on the other hand, women pursued a more successful and bountiful subsistence production. Women found, in market gardening, poultry raising, and dairying, important sources of household income when surpluses were sold in local markets. Many of these activities became capital-intensive industries in their own right by the late nineteenth century.[55] Unlike in Upper Canada, where women came to town to market the surplus of their gardens, many reports of Newfoundland women have them doing something quite different. As one Carbonear report typical of many in the 1830s to 1850s described it, 'On Saturday and Monday last our streets presented a melancholy appearance, about, we suppose, two hundred poor females came from the North Shore to seek relief from the Benevolent Irish Society of this town, which Society had given Fifty Pounds for the Poor, and these poor creatures had scarcely a garment to cover their nakedness.'[56] Similarly, a Harbour Grace newspaper reported in 1847 that it was not enough that young men could hope to find their subsistence in the seal fishery: 'Who, for the most part, will be left behind? Women, the aged, the decrepit, and the helpless! How are these to be fed?'[57] To meet the inadequacies of local subsistence agriculture, the Harbour Grace *Weekly Herald* recommended that women stop going to the merchants' stores to buy food, and think about ways to make fish offal part of their family's diet.[58]

Like women throughout the preindustrial Anglo-American world, women in nineteenth-century, northeast-coast fishing households were responsible for most subsistence production, particularly cooking, maintaining the family's clothing, and gardening. The traditional Newfoundland household economy, however, survived well into the twentieth century, whereas such subsistence production elsewhere was curtailed by the increasing pervasiveness of industrial-capitalist industries, many of which grew out of women's early subsistence activities. While twentieth-century families in Ontario obtained most of their subsistence goods from the purchase of commodities in the domestic market, families in Newfoundland continued to live by the merging of subsistence production and the fishery.[59] Women continued in their roles as shore-crew skippers, gardeners, keepers of livestock, menders of clothing, cooks, and chief family reproducers.[60] The cod fishery, requiring the labour of all household members, absorbed much of the female labour which, by developing local resources for the household's maintenance, provided important early domestic production for Upper Canada's fledgling local market. The resource endowment of the northeast coast proved to be a severe restraint on the number of subsistence activities women could engage in. A symbiotic relationship between the fishery and the household's other production (that of both men and women) continued as a strategy for dealing with the dominance of merchant credit in staple export. If families did occasionally generate surpluses from subsistence production, these likely ended up on the fish merchant's books rather than circulating in local exchange.[61]

The merger of women's household labour in market-oriented work in the fishery did give women an important place in northeast-coast society. Their crucial role in the making of salt fish tested, at least informally, the household's patriarchal structure. Their often violent defence of their households' relatively meagre subsistence resources was an additional challenge by women to female subordination in society. Yet the struggle for survival ensured women's full support for their households in the end, even though the latter were at heart patriarchal structures sometimes characterized by male violence. It may well have been that, because the courts allowed families much more freedom to discipline relatives forcibly, planters considered such family members a much more attractive labour supply than that hired on wages carefully regulated by the wage law of the time.

PART THREE: FISHING PEOPLE AND MERCHANTS

5

The Legal Regime of the Fishery

Beat them, suggested Sir Hugh Palliser in 1793, and servants would work harder and better for their planters, ensuring a more profitable fishery. The problem, the ex-governor of Newfoundland acknowledged, was that the act named for him in 1775 gave fishing servants freedom from discipline and guarantees for wages, which were extraordinary in the early days of British industrial capitalism. As a misguided attempt to force servants to return to Great Britain, the act forced planters to pay servants' wages without any effective way to exact penalties for negligence. Mandatory prearranged employment contracts forced employers to negotiate wage rates without reference to planter earnings in any particular season. The law must change, Palliser suggested, if the fishery was to prosper; he recommended a return to pre-1775 methods of disciplining servants by fines and corporal punishment.[1]

Planters found it easier to use labour subject to noneconomic forms of discipline than to hire servants protected by Palliser's Act. Fishermen generally resorted to family labour in household production with the support of merchants and the state. Family labour lowered the overhead costs of labour, minimized the risk of credit overextension, and alleviated the expenditure burden of government relief. In the context of a patriarchal social and legal environment it also provided fishermen with female and child labour easily disciplined by the fist, without the need to consider the legal complexities of violating wage contracts. The fishery could not, however, completely do without servants. Planters had to hire labour at points in a family's life cycle when the households could not generate enough of its own labour. Larger structural changes, such as those associated with the Napoleonic Wars, could also encourage planters to use more wage labour.

The wage lien posed just such an obstacle to planters who wished to hire labour, because of the manner in which local courts chose to interpret it. Section 17 of Palliser's Act actually provided for the imprisonment and whipping of deserting servants, as well as the forfeiture of all wages due for the time of contracted service (see appendix A). Such provisions were in keeping with the criminalization of servants' breach of contract in the Anglo-American law of master and servant, but local courts chose to ignore them. They preferred to emphasize the contractual obligations of masters to pay wages as a way of providing servants with the means to return to Great Britain. The Harbour Grace courts, for example, never used imprisonment or whipping to discipline negligent or deserting servants. Servants almost always won suits for wages against evasive masters in the Surrogate Courts and Courts of Session. Although they were less likely to win cases against masters who deducted wages for negligence, servants still won between 30 and 40 per cent of such suits between 1787 and 1799. Surrogates and magistrates, even in cases lost by servants, rarely allowed masters' deductions for negligence to stand without some arbitration in favour of their employees. While the Board of Trade never designed Palliser's Act to be a defence of servants' rights to the payment of wages, that was how local courts interpreted the legislation. The administration of wage law on the northeast coast protected servants more than their masters, a dramatic reversal of the manner in which the master-servant law elsewhere specifically favoured employers in wage suits, and more generally privileged them over their hired labour.[2]

After 1815, merchants, like their planter-clients, grew nervous about the use of credit to pay servants' wages. Merchants' worries, however, stemmed not only from uncertain economic conditions but also from the regulation of wages and credit in Palliser's Act. Although originally designed to provide migratory fishing servants with wage security, by 1802 the act had extended to resident servants, giving them preferential wage liens in the event of planter insolvency. Rather than supporting the development of an indigenous Newfoundland capitalist planter class, the wage and lien system smothered the efforts of planters to accumulate capital by allowing them little leeway in the contracts they formed with hired servants. Such legal inflexibility discouraged merchants from extending credit to planters who hired much wage labour. Merchants further refused to give credit to planters based on their long-term success in the fishery; they relied instead on the law of current supply, which gave suppliers of credit to planters the second lien on planters' fish as security for credit for the current season only. Not only did the law discourage planters from hiring labour, but by giving the greatest security to the shortest-term creditors, it also deterred merchants from making long-term investments in planter production.

Fish merchants immediately responded to Palliser's Act by arguing that the provisions for the return of servants to Great Britain restrained the development of many different kinds of service required by a complex and volatile trade. Furthermore, they said, the act's mild provisions for disciplining servants would make planters less likely to enjoy successful fishing seasons.[3] While merchants tended to blame all planter misfortunes on negligent servants, Chief Justice John Reeves observed in 1791 that planters suffered the most under Palliser's Act because it restricted their relationships with both servants and merchants. Reeves noticed that the act assumed that merchants directly manufactured salt cod. This was not so. Planters dealt with the problems of producing fish, while merchants concentrated on the profits to be made by supplying the trade in return for fish. Merchants charged such high prices for supplies and gave such low prices for fish and oil that planters earned little. Some merchants might not even issue supplies if they thought planters might not catch enough fish to pay for them. In the midst of the varying availability of merchant credit, planters faced the constant demand of servants' guaranteed pre-fixed wages; in a bad year, planters' ruin often resulted from the twin demands of merchant credit and servants' wages.[4]

Reeves sympathized with the merchants. The entire cod-fishery depended on merchant credit, since planters alone did not have enough capital to engage in the fishery as independents. They had to have merchant credit for supplies, but had only their catch to offer as security, and often tried to renege on their credit obligations by trading fish for even more supplies from some other merchant. If a planter failed, then the merchant had to pay the wages of the planter's servants over whom he exercised no form of discipline. Palliser's Act protected the servant, but not the planter or the merchant.

His hands tied by the imperial government's commitment to the letter of the act, Reeves could only recommend that naval surrogate judges arbitrate disputes between servants, planters, and merchants as impartially as possible under the appellant authority of the Supreme Court.[5] Both Surrogate and Supreme Courts continued to apply the

act's provisions, although the inflationary early days of the Napoleonic
Wars more than doubled the return fares to Great Britain, preventing
seamen and fishermen from leaving Newfoundland at the end of the
fishing season. Consequently, officials at Newfoundland began to over-
look the act's requirement that servants return home. Servants could
stay in Newfoundland and live off half their wages while their families
back home enjoyed the use of the rest.[6]

Meanwhile the 1793 British House of Commons select committee
investigating the judicial reform of the Newfoundland trade was gathering a great deal of evidence about how Palliser's Act discouraged
planters' employment of servants. It heard from Palliser himself, but
also from William Knox, a 'Newfoundland adventurer' and former
undersecretary of state, William Newman and Peter Ougier of Dartmouth, and a Mr Jeffrey of Poole, the last three representing the West
Country merchants. In addition the select committee took evidence
from Mr Graham, secretary to the Newfoundland governors, and Mr
Routh, the Newfoundland customs collector. Newman stated that the
planters faced a 'certain loss' in the fishery because servants learned
that the penalties of Palliser's Act regarding negligence prohibited masters from effectively disciplining their hired labour, or even dismissing
them, except in case of desertion. The act imposed only small penalties
on servants for poor work in an industry in which employers had
few means of controlling production during a restricted catching and
curing season. The witnesses agreed with Newman's claim that the
wage and lien provisions of Palliser's Act ruined many planters because
servants worked hard enough only to cover their own wages, not to
ensure the planters' profit.[7]

All witnesses had reported that Palliser's Act did nothing to stem
the growth of the resident fishery, yet the act continued. Throughout
the Napoleonic Wars it remained the guiding force behind wage law
in the Newfoundland cod trade. The wars themselves, however, by
disrupting the migratory fishery, overrode the provisions within the
act that were designed to secure the return of servants to England.
While naval governors perfunctorily attempted to enforce Palliser's Act
throughout the war, officials at Newfoundland and London increasingly accepted that the fishery had become a resident one.[8]

A ruling by the Supreme Court confirmed that the wage and lien
guarantees of Palliser's Act would survive in the resident fishery, but
without provisions for servants' return to Great Britain. In 1802, for
example, Justices Tremlett, Williams, and Cooke debated whether or

not the penalties of Palliser's Act applied to one Halfyard, who had advanced his servant, Thomas Coysh, almost all of his wages during the fishing season without reserving the usual half. Coysh was originally from England, but had married and taken up residence at Newfoundland. While Williams wanted to enforce the exact letter of Palliser's Act, Tremlett and Cooke disagreed, arguing that the act applied only to migratory seamen and fishermen. Servants could now take up residence in Newfoundland without having to worry about their masters sending half their wages to Great Britain. Moreover, under the regulations of the Judicature Act, resident servants received the same rights to attach for wages the goods, debts, and effects of their masters. The Judicature Act especially applied to insolvency cases, and contained no provisions for the return of servants to Great Britain.[9]

The Judicature Act and this Supreme Court decision carried the wage and lien system over, from the law governing the migratory fishery to the law governing the resident fishery. While the wage and lien system protected servants unusually well, servants did not enjoy either an exploitation-free existence or even an easy one. During the Napoleonic Wars servants sometimes voluntarily accepted service in the British army or navy when they met up with the many naval and army press gangs that roamed the northeast coast (also to evade debts they had incurred in the fishery). It is a telling comment on the harsh conditions of employment in the Newfoundland fishery that the Newfoundland governors between 1802 and 1807 – first Gambier and then Gower – prohibited by decree press gangs accepting servants who had unfulfilled agreements with planters. Press-gang recruitment deprived masters of the labour they needed to conduct the next spring's seal fishery. The governors ordered the gangs not to take any servants without their masters' permission, because the former had taken credit for winter and the latter had guaranteed their service in the seal fishery. Without that service masters would lose the voyage, and yet be responsible for servants' supplies which had been purchased by the master on credit.[10]

The governors' regulations, by recognizing that resident planters needed servants year-round, acknowledged implicitly that the migratory fishery at Newfoundland was dead; that merchants conducted all of their business with resident families. However, the wage and lien system continued despite this change. The legal infrastructure inherited from the migratory fishery and entrenched in the wage and lien system discouraged planters from relying more on the use of wage labour in production. Meanwhile those who wanted to remain solvent retreated into household production through family based labour and dependence on merchant capital.[11]

Governor Gower therefore recommended the removal of all regulations which impeded the resident fishery, especially Palliser's Act; yet the Colonial Office engaged in no such legislative reform. On its recommendation Parliament instead enacted the permanent Judicature Act of 1809 (49 Geo. III, c.27), which continued the rights of current suppliers and servants.[12] Suppliers of capital goods, provisions, and clothing for the current fishing season did not have to fear previous years' creditors seizing the catches which had been made on the labour or credit of current suppliers'. Current suppliers could sue for the sale of their debtors' boats and other property in the Supreme or Surrogate Courts. But if a current supplier allowed an indebted planter to continue to obtain credit for the next season, he then lost current supplier status and could not take action to seize the effects of a debtor – if such action jeopardized the security of payment to the new current suppliers.[13]

The supplying merchant generally paid the wages of servants through credit on the hiring planters' accounts, in which the planters' fish and oil, caught by servants, was posted against servants' debts throughout the fishing season. If a planter failed, the claims of his servants took precedence over those of the merchant under the lien of current supply. This angered merchants, who felt that those who invested capital in the trade through extending credit to planters bore the greatest risk. In 1809 the Society of Merchants at St John's demanded that Parliament amend the new Judicature Act so that, in the case of planter insolvency, the claims of current suppliers, and then all other suppliers, received preference over those of servants in the settling of insolvent planters' estates.[14]

Chief Justice Tremlett incensed merchants by refusing to allow them to engineer, through the courts, a hybrid enforcement of both Judicature and Palliser's Acts. The merchants wanted their preferred status as current suppliers of credit extended to claims against wages. They argued that they indirectly supplied servants by advancing credit to planters (who later became insolvent) to pay wages. Consequently, insolvency law should give merchants the most preferred status, because without their credit no wages could be paid in the first place. Tremlett disagreed. He argued that, although the fishery had been altered by residency, Palliser's Act still provided that servants employed by planters

must have one half of their wages reserved for payment at the end
of the fishing season. Servants could charge more than their entire
wages to their own accounts or to those of their planters with merchants during the fishing season. But now, servants' remaining half-
wages were safe from the merchants' lien for credit they took up during
the fishing season. Palliser's Act, by insisting that nothing interfere
with the payment of the reserved half-wages, prohibited merchants
from retaining planters' fish and oil which had to be sold so that the
proceeds would allow planters to pay their servants. Merchants did
not feel that servants should have the first claim on a planter's voyage
if the servants owed money themselves. But rather that merchants
should be able to seize fish and oil directly from planters to balance
against servants' debt. Chief Justice Tremlett would allow no contract
for debt entered into by servants to infringe on the security of their
wages.[15]

To add insult to injury the chief justice further argued that merchants had to honour all bills of exchange that their clients might
draw on them to pay wages, regardless of the success or failure of
planters' voyages. Palliser's Act guaranteed that wages would be paid
in good bills of exchange, not ones returned because the master who
issued the bill had a poor credit standing with his merchant. Tremlett
ruled that only planters and their supplying merchants were in a position to know exactly how solvent the planters were. Neither planter
nor merchant could expect servants to have access to such knowledge
nor could they take any action which could violate the right of a servant
to payment of wages. Planters and merchants were the only ones who
knew how much fish had been shipped throughout the fishing season.
Servants did not have access to such information. As a result, planters
who issued bills to pay servants' wages, knowing that their voyage
had not realized enough fish and oil to meet them, had to find other
means of paying. Issuing bad bills of exchange, however, did not free
planters from the servants' lien on the fish and oil of the voyage on
which they were employed.[16]

Merchants responded that the fishery was no longer well served
by Palliser's Act, and that Tremlett's decisions should not be based
on an obsolete law. Furthermore, the root of the evil in merchants'
eyes was the way servants seemed to get the best of the legal system.
Planters did not always issue bills to defraud servants intentionally,
but they would sometimes become insolvent before servants could
send bills to the home firms of the English partners. The merchants

saw no justice in Tremlett's refusal to attach servants' wages for debts. The whole of Palliser's Act no longer applied fairly to the payment of half-wages in bills of exchange because few servants returned to Great Britain. Rather, they continued to stay in Newfoundland, take goods on credit during the winter, and find employment in the spring seal fishery. The whole transatlantic system of payment should have been struck down in favour of the truck practices merchants used with resident fishermen. It took time to negotiate bills in Great Britain for servants residing at Newfoundland. Servants took credit against those bills, and should have been liable to legal process if they did not honour their accounts, even if the drawer of their bills for wages became insolvent. Wage law designed for the return of servants to Great Britain in a migratory fishery had no place in the administration of a resident one. Palliser's Act consequently had to fall away.[17]

Tremlett's refusal to compromise with merchants on wage law led the Board of Trade to recommend the appointment of a new chief justice who would be open to a wider interpretation of the laws at Newfoundland. Yet believing that the wars' end might re-establish the migratory fishery, the British government at first took no steps to remove the guarantees Palliser's Act afforded to servants for their wages. In 1817, however, when Colonial Office administrators became aware that a gap existed between the intentions of Newfoundland legislation and the reality of the resident fishery, they continued to ask the governors whether or not the Judicature Act should be reformed. Although the Newfoundland courts accepted that the wage lien protected resident servants, the law still declared officially that only servants from Great Britain or Ireland had first lien for the payment of their wages in any current season. Governor Pickmore recommended that the law be amended officially to give all servants including residents and not only those from Great Britain and Ireland – a lien on the fish and oil of an insolvent planter. After the servants were paid, the current suppliers of credit for goods absolutely required by fishery would be paid, followed by all other current suppliers, and finally, all other creditors, rateably, out of any remaining effects. If the fish and oil were not sufficient to cover seamen's and fishermen's wages, these men might also share rateably in the remaining effects of the insolvent planters. Pickmore agreed that no planter should collude with a merchant to sell fish or oil during the season without the knowledge of the servants. If an insolvent estate revealed that merchants bought and sold fish without regard for the security of the servants who

Extend protection!

caught the fish, then such buyers and sellers were liable for a fine of one pound per quintal of fish and £20 per ton of oil sold.[18]

Post-war depression focused attention on the manner in which the wage and lien system exacerbated the inability of planters to achieve any profit through the use of wage labour. J. Newart wrote to the secretary of state for the colonies in 1817 to state the case for the island's planters. Newart was angered by what he felt was an injustice – that planters took responsibility for actually producing salt fish and bore most of the trade's risks. Yet wage law unduly benefitted their servants to the prejudice of planters. Planters had to pay wages according to their contracts regardless of the quality of the servants' work. Newart further argued that planters could not make consistent profits in the fishery because the law compelled them to sign agreements with servants fixing wages before the season began. Servants shared none of the voyage's risk, either in highly variable catches or in market prices. Furthermore, the law compelled planters to pay wages without any real means for making deductions for negligence or insubordination.[19]

Merchants were the bigger villains in Newart's analysis because their control over credit allowed them to fix prices for everything they sold to, and bought from, planters, in order to ensure their own profit at the end of a fishing season. Planters could not hope to survive under the burden of truck.[20] The great injustice of all this for Newart was that planters could not do the same to their servants. Seamen, fishermen, and other fishing servants had the right to be paid the full amount of their wages before anyone else received any money from the sale of the planter's catch. Servants also had the right to follow the fish and oil into the hands of supplying merchants. The combined forces of the wage and lien system and truck in the Newfoundland fishery created a pincer movement which forced planters into poverty. Merchants, because they had no way to proceed legally against servants' wages for debt, took their planters' fish to pay servants' debts. Planters then had to recover these sums from their hired labour on their own.[21]

The legal regime of the fishery encouraged planter insolvency. According to Newart, the immediate aftermath of the wars saw some merchant houses fail, but it was primarily planters who either went broke or were forced to reduce the scale of their operations greatly, before they became insolvent. Planters could only survive and prosper if all the laws generated in the interest of the migratory fishery disappeared. Merchants who had formerly opposed the wage and lien system had found a way to use it to their advantage. Planters were effectively obligated to their merchants' credit for both supplies and servants' 'extravagant' wages. If they could not pay all their debts to both merchants and servants, planters lost all their property to merchants who seized fishing rooms, plantations, dwellings, and equipment to pay the debts.[22]

Newart suggested that Newfoundland needed a better judicial system if planters were to succeed. He advocated clearing away the underpinnings of merchant domination of the Newfoundland fishery – the wage and lien system – so that true capitalist competition could assert itself in the fishery. Merchants would then only extend capital to successful planters. Planters would succeed because, no longer hampered by Palliser's Act, they could more severely discipline servants and renegotiate wages to better reflect the success or failure of a fishing voyage. Planters and servants who did not succeed would have to hire themselves to solvent, prospering planters on shares so that they could only make claims to a reasonable proportion of a planter's actual voyage.[23]

The 1817 Parliamentary Select Committee investigating the provisions crisis heard similar condemnations of the wage and lien system. Poole merchant George Garland testified that West Country merchants accepted that the resident fishery was the most profitable way they could obtain their staple commodities. He argued that servants and current suppliers should have an equal preference in claims on the produce of the planters' voyage. The pincers' pressure should balance, but not lessen. Planters found no relief in Garland's recommendations. James Henry Attwood, who represented St John's merchants, agreed with Garland. Attwood was offended by the manner in which Palliser's Act insinuated itself into the shaping of the Judicature Act. The act now allowed servants the preferred claim for their wages out of the entire estate of insolvent planters, not only from the fish and oil.[24]

Despite all of the debate about the exact workings of the wage and lien system, most commentators agreed that it did not help planters. To hire labour, planters had to accept wages fixed by written contract before they could possibly know how the season might fare. The law did not permit a master to ask his servants to accept a wage reduction or force a wage roll-back so that he might survive the fishing season solvent. Furthermore, even if they caught and cured a good voyage, the law of current supply limited planters' freedom to market. If a planter tried to send his fish to anyone else, his current supplier of credit might become nervous that he might lose his preferential lien, go to court to attach the planter's voyage, and possibly force the planter to ask to be declared insolvent. If a planter wanted to be sure to avoid prosecution by his merchant, he had to continue to deal with him regardless of the prices the merchant charged for supplies or gave him for fish. Current suppliers additionally guarded their status as preferred creditors because they envied the even stronger claim allowed to servants by law. Finally, merchants had little reason to offer better fish prices to planters who were not their regular clients, because as the most unsecured creditors they stood to lose if a planter's current supplier forced an insolvency through legal action. Planters were stuck between a rock and a hard place, between servants and current suppliers, in which there was little room for their own success.

Supreme Court decisions came to reflect the growing uneasiness with the manner in which the wage and lien system was hurting planters. On 23 October 1817 Chief Justice Forbes ruled, in the case of Crawford and Company versus Cunningham, Bell and Company, that the definition of current suppliers had taken the lien provisions of the Judicature Act too far. He stated that Palliser's Act established current supply provisions when merchants transported fishermen of little means to catch and cure fish at Newfoundland. These fishermen could only offer their prospective earnings as security for credit. Forbes felt that it was 'natural' for a custom to develop whereby fishermen and merchants had preferable claims for wages and credit from migratory boat keepers. However, as settlement increased, this custom outgrew its usefulness and became a problem through its entrenchment in the Judicature Act. Forbes felt that the true intention of the act was to make a more equitable division of property among all creditors than was allowed by current supply. But the Judicature Act departed from conventional insolvency law to reflect the unique requirements of fish-producing. Planters did not have any capital or credit, 'except such as they could raise upon the fish they might catch in the season,' to guarantee servants' wages or merchants' credit. The law consequently tied planters to their supplying merchants through the preferable claim of current supply, after merchants first satisfied servants' wages to the fullest extent that the planters' catches allowed.[25]

Forbes argued that the impoverishment of planters, which made such a credit system necessary, meant that planters could not be considered independent employers. The working of the wage and lien system was a *de facto* recognition by the law that planters were little more than middlemen between servants and merchants. Planters guaranteed wages based on faith in their supplying merchants' ability to pay, not their own, which was why merchants, not planters, honoured the bills of exchange used to pay servants' wages. Merchants funded the fishing voyages, bore all the expenses of marketing the fish, and ultimately paid servants' wages. For these reasons alone, servants had the right to follow the fish and oil into the merchants' hands. Ultimately, the solvency of a planter rested on the solvency of his merchant, so the merchant was the true employer of a planter's servants. Planters did not have the resources to hire servants on their own, but only by the special credit props of the law of current supply.[26]

Yet merchants firmly believed that the preference given to wage claims on the proceeds of a fishing voyage jeopardized the credit they extended to planters. Forbes ruled that merchants had a right to organize production by insisting that planters hire servants on shares rather than by a guaranteed, pre-set wage, because it was in fact the merchants' capital which truly employed servants.[27] Sharemen would only receive a return from the planter in proportion to their actual productivity during the voyage. In the case of Stuarts and Rennie versus David Walsh, the plaintiffs alleged that they had supplied two planters, Merigan and Jervis, on the sole condition that they ship servants to the Labrador fishery on shares alone, and that the credit issued by Stuarts and Rennie would be paid before any wages. Merigan and Jervis's voyage failed, not allowing them to pay for supplies. Walsh, a servant of the two planters, received his wages. Stuarts and Rennie argued that Walsh, as a shareman, shared in the responsibility of the credit. Forbes accepted Stuarts and Rennie's argument, recognizing that, in response to the inflexibility of the wage and lien system, merchants and planters increasingly would hire servants on shares, not wages. In Forbes's opinion, the wage and lien system was itself destroying the employment of wage labour by planters in the fishery. Chief Justice Forbes declared Palliser's Act obsolete. Merchants could finally attach shares, unlike fixed wages, in planters' hands for debt.[28]

At the same time as the chief justice was trying to use the share system as a means to circumvent the wage and lien system, some of the naval surrogate judges tried to renew enforcement of Palliser's Act.[29] Captain Nicholas of HMS *Egeria*, patrolling Trinity Bay in 1820, decided that masters must observe the conditions of the act governing employment of servants. But Governor Hamilton wrote that the exact enforcement of the wage lien would hurt 'Capitalists embarked in the trade and fishery of this Island.' In his opinion, merchants and masters, now that the fishery was a resident one, were no longer obliged to guarantee wages under an act expressly designed to secure the return of servants to the British Isles. Masters were obliged to pay servants the 40 shillings formerly reserved for their passage home, but that exhausted the obligations between the two. Chief Justice Forbes might have felt that Palliser's Act was dead, but Nicholas maintained that as long as the act remained on the books he was going to observe it in his rulings.[30]

Nicholas's intransigence in protecting servants' wages appears to have stemmed from his desire to see somebody other than merchants benefit from the fishery. He was absolutely dismayed to find the industry dominated by indigent planters who were completely dependent on merchant capital, family labour, and at times, labour hired on shares. Nicholas suggested that planters who risked hiring labour were least likely to be able to repay their debts. In such cases merchants were quick to take legal action, which forced insolvency before they lost their status as current suppliers. The captain remarked that the planters and servants of Newfoundland 'really appear to me to be more like the slaves of a feudal lord, than the free subjects of a Great Nation.'[31]

The willingness of surrogates like Nicholas to uphold a law considered obsolete began to add fuel to demands for legal change at Newfoundland by the St John's reformers. In 1821 Governor Hamilton asked the Colonial Office to appoint an attorney-general for the island, pointing out that naval officers were ill-trained to handle matters of civil jurisdiction.[32] Reformers partially demanded legal change because they felt that naval surrogates arbitrarily substituted naval discipline for justice, as was the case when surrogates Buchan and Leigh had planters Butler and Lundrigan whipped in 1820.[33]

The disputes between merchants and the Surrogate Courts over the application of wage law continued to grow. Surrogate Courts continued to rule that servants had the right to a prearranged, fixed wage as stipulated by Palliser's Act. Merchants complained that, as most servants were now hired on shares rather than fixed wages, the surrogates must accept new arrangements in the fishery. If not, merchants would either fail or withdraw from the trade. Either way, this would leave servants and planters at Newfoundland without the means of subsistence, let alone the capital to engage in a fishing voyage.[34]

By 1822 officials in the Colonial Office had decided that the Judicature Act allowed surrogates too much influence over the commercial transactions of the fishery, and that Palliser's Act should be
replaced. Servants now resided in Newfoundland, so there was no
use in having wage guarantees that secured their return to Great Britain. Furthermore, the depressed fish trade could no longer support
such guarantees. Thus, it made sense that merchants should control
planters' employment of servants by being mandatory parties to all
contracts between the two; also that servants should be allowed no
preferential claim on the estate of insolvent planters because they were
just as much dependent on the current supplier as was the planter.
Frequently servants' 'misconduct' caused the planters' failure as well.
Although willing to sponsor wage-law reform, officials were only prepared to defend the law of current supply as an essential security to
the merchants' capital advanced on credit to fishing people.[35]

Merchants in the Newfoundland trade agreed that the provisions
of the Judicature Act allowing servants first lien on the planters' voyage
had to be struck down in the interest of labour discipline. Thomas
Hunt wrote to William Newman, a merchant at Dartmouth, that
changes in the act giving servants no more than an equal claim to
the current supplier would force them to work harder in the fishery,
thus insuring that planters would have more successful fishing seasons.
James Dutton of Liverpool also lost no time in letting the Colonial
Office know that the laws governing servants' wages must change.
Dutton suggested that a new law fix a minimum catch which would
pay for servants' outfits and wages, that if the planters' total catch
was not sufficient, servants' wages should be reduced. As an example
Dutton pointed out that a schooner sailing to Labrador with ten hands
usually made a voyage that would pay servants' wages and provisions
if it brought home 1,000 quintals of fish. Therefore, if a schooner
caught only 900 quintals in a season, the law should allow planters
to reduce servants' wages by one tenth. Such a change would allow
planters to calculate wages in light of the prices of provisions and
fish. In addition, servants would work harder in this kind of piece
system and would also watch carefully that planters delivered all their
fish to the current supplier or 'lose part of their wages the Merchant
being the person who pays them.'[36]

In 1823 the Colonial Office, proposed replacing all previous legislation governing the Newfoundland fishery with a new Judicature
Act. The new act would prepare the way for the granting of repre-

sentative government. Inferior district circuit courts would replace the Surrogate Courts in the outports. Insolvency regulations would continue to allow servants first claim on the estate of planters for wages. Current suppliers would have the second lien. The new law would confine servants' claims to fish and oil only. Employers could advance all but a fourth of the wages in goods throughout the fishing season. Fishermen absenting themselves from work could be penalized five days' wages for every one day missed.[37] Governor Cochrane arrived in 1824 armed with the new Judicature Act (5 Geo. IV, c. 67). This act began the new circuit court system, which would try all civil disputes according to English law and custom.[38]

A new Fisheries Act accompanied the 1824 Judicature Act, supplanting Palliser's Act and any other laws governing the fisheries. The Fisheries Act (5 Geo. IV, c. 51) explicitly recognized the resident fishery. While continuing to insist on written contracts between masters and servants prior to the start of a fishing voyage, the new act recognized both wages and shares as legitimate means by which masters could pay servants; it also allowed planters to advance goods on credit against three-quarters of their wages. Servants still had a lien which allowed them to follow fish and oil into the hands of the merchants, but the lien now explicitly secured shares as well as fixed wages. The Colonial Office meant the act to be a temporary expedient designed to disrupt the resident fishery as little as possible, until such time as a colonial government could take over responsibility for the regulation of the Newfoundland fishery.[39] The Judicature and Fisheries Acts of 1824 superficially appeared to reaffirm the wage and lien system. But by sanctioning the right of masters to pay their servants by shares – and extending lien protection to shares – the legislation simply reflected the triumph of shares over fixed wages as the means of payment favoured by planters.

Administrators in the Colonial Office hoped that merchants and planters would gradually become involved in fewer court cases with servants because of the increased costs of the new circuit courts. Both laws continued to recognize servants' liens because officials believed that planters, stuck between their obligations to both servants and merchants, were rarely solvent. Colonial officials wanted to give merchants first lien, before servants received their wages, because they believed that merchants risked the most in the trade by advancing the credit on which the fishery rested. In the end, the acts simply made the restriction of servants' liens explicit to the fish and oil they actually produced. Servants could no longer hope to receive payment from the sale of insolvent planters' other effects.[40]

The advantage of this change to both merchants and planters was that, if a planter's voyage failed and there was not enough caught to pay wages, servants could hold neither planters nor merchants responsible for any agreements made at the beginning of the season.[41] According to the logic of the new act, neither current supplier nor planter was responsible in any way for servants' earnings beyond what the latter caught. The Colonial Office did not want to remove the lien altogether because it noted a disturbing fact of the share system: fishermen who earned nothing from the fishery still had to eat. Unless they had some rights to earnings, fishermen could well become a constant burden on the public purse for relief. The Colonial Office insisted on a balance between its fiscal concerns and the desires of planters and merchants.[42] The new Judicature and Fisheries Acts served their purpose in allowing the Colonial Office to shift responsibility for winding down the legislative confusion surrounding the wage and lien system. The extended life of these temporary acts cleared the Colonial Office of any further responsibility for legislating in such matters until the Crown finally granted Newfoundland representative government in 1832.[43]

There is little evidence to suggest that the laws governing the wage and lien system gave any encouragement to the formation of industrial-capitalist relations between planters and their servants in the first half of the nineteenth century. Such laws, beginning with Palliser's Act of 1775, simply made planters more vulnerable to failure. Palliser's Act, an arm of restraint on planters which originated in the migratory fishery, gave servants a protection far beyond the ability of planters to pay and still make a profit in the fishery. In the early resident fishery, wage labour did not prove to be a variable cost which planters might manipulate to their advantage.

The special lien of current supply given to merchants did ensure that planters had access to the credit they required to provision a fishing voyage. But the law of current supply meant that planters had to offer their fish to supplying merchants first, regardless of the price offered. Acting as it did in concert with truck, this made it difficult for planters ever to accumulate much capital in their own right. The guarantee of a set wage to servants pre-fixed before the season even began seriously exacerbated the planters' problems. Already restrained by the costs of merchant credit, planters could not renegotiate wages to suit shortfalls in either the catch or the prices of salt cod and fish oil. The wage and lien system allowed merchants the right to control the organization of a planter's production. Both planters and merchants cooperated in resorting to the share system as a means of side-stepping the worst impediments of the wage guarantees. Planters, when they could hire servants, remained the merchants' middlemen, paying shares and using credit as operated by merchant capital. Like other fishermen they continued to rely on family labour, except during times when the family could not provide all of the household's requirements. The wage and lien system did not alter this trend towards the family fishery, but rather proved to discourage planters' experiments in the accumulation of capital through the use of wage labour.

6.

Truck as Paternal Accommodation

Overpowered and confined by men with blackened faces at Brigus in 1848, bailiff William Lilly heard his captors declare '"Bowring you son of a bitch we will make your Goods pay for it."'[1] Lilly had been guarding the premises and goods of local fish merchant Richard Leamon, recently attached for payment of debts owed to his suppliers, Bowrings of St John's, by the Northern Circuit Court at Harbour Grace. Lilly's assailants, although never discovered, were probably local planters and fishermen who determined that one of their own in Brigus, a merchant whom they counted on for supplies, credit, and markets, was not going to be eaten up by some big-time merchant outsider from St John's. Their solution was simple: steal the goods remaining in Leamon's store before the court could sell them to satisfy Bowrings' claim.

The Brigus incident suggests that the social relationships which developed between merchants, planters, and other fishing people were diverse and complex beyond the prescriptions of wage and credit law. Statutes established the limits of what the state would permit in the relations between such people, but there was much room for popular negotiation and adaptation before disputes might end up in court. The judiciary arbitrated this process according to the legal regime of the fishery only when it broke down. The evidence gathered by the courts, nonetheless, is a useful source which hints at the ways in which fishing people and merchants confronted each other before they resorted to state intervention. While the discourse of law may have fixed broad parameters, a very material class struggle governed the day-to-day lives of residents of the northeast coast.

The actions of the Leamon burglars indicate that the outcomes of such struggles were sometimes ambiguous. Did they commit the deed for Leamon or for themselves? Did they feel Leamon to be someone worth defending – he was, after all, a fish merchant – or did they simply detest the Bowrings more? The equivocal nature of the relationship between Leamon and the fishing people of Brigus lay rooted in credit, which more generally served as the nexus between household producers and merchants on the northeast coast, and which defined class relationships between fish merchants and fishing families (from the most propertyless to those of planters with schooners in the Labrador and seal fisheries). Truck represented a mutual, though unequal, accommodation between two basic classes: merchants and fish producers. The need of producers for credit to purchase capital, consumer goods, and labour (particularly during early settlement) dominated the Newfoundland fisheries.[2] Although this need was constant, producers faced frequent, cyclical depressions in the industry due to wars and variations in market demand and supply. A planter might be exploited by his merchant's price manipulations, but the merchant 'at least ... kept him alive.' Merchants, in turn, could increase the prices of goods they sold residents to offset losses in the fish trade.[3]

Fish producers and merchants, like suppliers of labour and capital in other staple industries, needed each other, but this interdependency did not preclude struggle between the two.[4] In other parts of British North America, the producing classes – whether servant, artisan or farmer – sometimes enforced the rule of accommodation by rough behaviour or riot when their dominant partners strayed beyond accepted limits of behaviour.[5] Paternalism in the fishery, the ideological expression of truck, was no simple merchants' tool used to ensure their hegemony. Merchants did not totally control their credit relationships with fish producers. They had to accommodate the separate purposes of their clients, who often took actions which ensured that truck continued to meet their needs. Many fishermen did not accept without challenge the manner in which both planters and merchants used truck to profit in the fishery. While such challenges, especially in the courts, did not end truck, they did limit the extent to which it exploited producers. Although merchant credit cemented cross-class ties between producers and fish merchants, it was the struggle between the two groups over its use which helped to define the character of northeast-coast society.

Court records reveal much about northeast-coast social relations during the first half of the nineteenth century. Debt disputes dominate the sample of writs issued by the Harbour Grace Northern Circuit Court from 1826 to 1855, constituting almost two-thirds, or 344 out of 542 writs (see appendix B). Unfortunately most of these writs identify little about the people involved in the court actions, and are consequently of little help in an examination of northeast-coast social relations. The sixty-eight disputes in the writ sample that specifically concern wages do allow some tentative conclusions about the experience of servants, planters, and merchants in the fishery. Most of the writs issued in wage disputes for Conception Bay involved fishermen. Only six involved servants not directly employed in the fishery.[6]

Seven of the wage-dispute writs concerned conflicts between servants in the spring seal fishery and their masters and merchants. Sealers usually did not sue their masters, but looked to the merchants who received their seals to pay their shares. On occasion these suits were not only for wages, but also represented servants trying to protect themselves from the worst parts of truck with merchants. Michael Patten, for example, sued merchant William Bennett in 1832 for £3 overcharges by the latter on his account.[7] Sealers kept accounts with their masters' merchants for needed equipment which they acquired on credit. Patrick Power, for example, received only £4 out of his share of £12 after Ridley, Harrison & Co. balanced his account in 1840 (see table 3).[8] Sealers were fishermen who simply engaged in a different industry for an extremely short period of time each year. Merchants tried to use truck in the seal fishery just as they did in the cod fishery.

The remaining fifty-seven wage-dispute writs all involved fishing servants. The patriarchal nature of social relations in the fishery emerges in some of these cases. While a very few women servants sued directly for their wages, most servants who used the courts to defend their wages were men.[9] In two cases fishermen appeared in court suing employers for both their own and their wives' wages. The suits suggest that planters kept accounts with their servants, advancing at least part of their wages as credit. The wages of both spouses appeared as one under the husband's credit. For example, in 1826 Peter Keefe sued Robert Knox for the balance of his wages: £19. Keefe's account with Knox, a planter, shows that the balance due was based on both his and his wife's wages (she remained unnamed in the document) (see table 4).[10] William Brennan's suit for £27 wages against planter Thomas Pynn of Musquito in 1827 contained a simple state-

TABLE 3
Account of Patrick Power with Ridley, Harrison & Co., 1840

Debits			Credits	
07 Feb	1 canvas frock	£0.06.0	7 May share seals	£11.03.9
10 Feb	2 yds. blanketing	0.08.0	less berth paid	
	3 1/2 yds. canvas	0.08.2	owner schr. *Emerald*	1.14.0
	1 pr. men's hose	0.03.6	Total	£ 9.09.9
13 Feb	1 pr. men's boots	1.10.0		
	1 pr. men's hose	0.03.6		
	1 ball hemp	0.01.0		
	thread	0.00.6		
17 Feb	1 flannel shirt	0.07.0		
	1/4 yd. whitney	0.03.0		
22 Feb	1 yd. blue calico	0.01.0		
26 Feb	1 pr. drawers	0.06.6		
	1 knife	0.01.6		
	1/2 yd. blanketing	0.02.0		
	1/2 yd. flannel	0.01.3		
02 March	1/4 yd. cloth	0.05.0		
	1 southwester	0.04.0		
	1 bowl	0.006		
	1 lb. leather	0.03.6		
	1/2 lb. tea	0.02.6		
	short paid on bread	0.10.0		
Total		£5.08.5		

Source: PANL, GN5/3/B/19, HGCR, box 42, file 1, writ no. 65

ment that the wages were for himself and his wife (again unnamed) the past summer.[11]

Servants on shares initiated only five wage disputes in the sample of writs. This low number suggests the greater ease planters and merchants found in paying a wage directly indexed to the season's catch, rather than paying a fixed sum set between planters and servants before the season began. Servants such as Jeremiah Pumphry sued for small accounts. Pumphry demanded only three pounds from his master, James Ball, in 1826. John Mugford did not even bother to sue for a cash value from his master, Charles Boon, in 1833. Mugford

TABLE 4
Account of Peter Keefe with Robert Knox, 1826

Debit		Credit	
Balance due from last yr.	£ 2.00.00	For work done as	
My diet	8.00.00	per a/c	£00.19.9
10 June 1/4 lb. thread	0.01.09	Overcharge on	
35 lb. soap	1.03.06	tobacco	00.01.6
1 pr. women's boots	0.11.00	Wife's wages	
12 Sept. 7 yd. bombazett	0.14.00	for summer	08.00.0
1 yd. calico	0.00.10	Keefe's summer	
6 June 1 1/8 yd. check	0.01.08	1/2 wages	23.00.0
10 June 3 lb. tobacco	0.004.06		
1 1/2 doz. pipes	0.00.06		
12 June cash	0.01.00		
cash	0.08.00		
Total	£13.00.10	1/2[a]	£32.01.3
Balance due Keefe	£19.00.04	1/2	

Source: PANL, GN5/3/B/19, HGCR, box 41, file 3, writ no. 241
[a]Mistake in addition is in the original.

simply demanded his share of the total amount of fish he caught: 12 quintals. At times sharemen could sue for large amounts, such as in the case of Francis Barrett versus John Barrett of Bishop's Cove in 1853, for £30 of the alleged value of his half-share of fish. But the advantage of the share system to employers emerges in the case of Thomas and Patrick Healey's suit against their master, James Walsh, for wages of £25 each in 1842. Walsh told the court he had not guaranteed the fishermen this amount, but instead had hired the Healeys on shares. The Northern Circuit Court found that the servants' share allowed them only thirty shillings.[12]

Although merchants and planters justified stripping away sharemen's protection under the wage and lien system because of Forbes's 1817 ruling that most had become co-adventurers with their planters, masters did not treat sharemen as equals. A number of actual contracts in other types of court documents suggest that, contrary to Supreme Court rulings, servants on shares were not co-adventurers with their masters. Sharemen were subordinate to their employers just as servants hired on fixed wages were. A case in point is that of James Pumphry, who at Carbonear on 5 June 1826, 'agreed and shipped myself to serve James Ball as a shareman from this date until the whole of the voyage is off. I am to have half my catch of fish after paying six pounds for my birth [*sic*]. Also to assist all in my power toward making the voyage when in from fishing the same as another man. I am to come home in the Schr. [schooner] the first trip if required.'[13] Other sharemen's shipping papers support viewing them as the employees of planters during the length of their agreements.[14]

The amounts sued for by fishermen not identified as sharemen in court records were much higher than those of sharemen. There were seven writs issued on behalf of fishing servants for less than £10. But only three – that of Thomas Melvin for £4 in 1827, Patrick Rogers for £3 in 1829, and William Walter for £8 in 1834 – suggest that the amounts sued for represented the total wage earned by servants. The other writs specified that the amounts sued for by servants were the balances of wages due. As in the case of the sealers, this suggests that planters kept accounts with their servants with which they advanced supplies on credit against wages.[15]

Planters could take advantage of their control over servants' accounts with them to maximize their profit in the supply of food, clothing, and equipment, which sometimes amounted to the half-wages allowed to be advanced by Palliser's Act. Servants had accounts with their planter masters, just as planters in turn had accounts with their supplying merchants. Other servants might well have a direct account with their masters' supplying merchants. No matter what the arrangement, wage law allowed no master to use credit to erode the half-wages balance due at the end of the fishing season. Planters, however, could try to manipulate the prices of goods supplied to servants so that it would appear they did not owe wages at the end of their contracts. A few cases indicating that planters used truck to avoid wage payments appeared in the Surrogate Court, which usually agreed with servants when the latter complained that planters overcharged prices on their accounts to avoid paying wages. In 1787, for example, Surrogate Packenham readjusted prices and ordered planters to pay wages then due to servants.[16]

Servants did not passively accept planters' or merchants' use of truck to undercut their wages, even before the establishment of the circuit courts. They used the Surrogate Court at Harbour Grace to ensure that their masters observed the letter of their prearranged wage and service agreements according to Palliser's Act. Many fishing servants who won their cases simply demanded that their masters pay their wages at the end of the fishing season, according to Palliser's Act, which the Surrogate Court judges then ordered. In the 1787 case of David Cushan's suit against his master, John Dowdle, for example, Dowdle made clear that he had not made enough from his fishing voyage to pay his servant's wages. The court ordered Dowdle's fishing boat sold to pay his debt. Masters who appeared in the Surrogate Court usually gave their servants half-wages in credit for required goods during the fishing season, but did not want to or were not able to pay outstanding balances for the rest at the season's end. Planters were more concerned to make sure that they satisfied the credit of their supplying merchants. For example, in 1787 Surrogate Packenham insisted that planters could not deal with merchants as if their servants' wages could wait until they met the supplying merchants' credit. Packenham ordered that the supplying merchants meet servants' wages before they could credit any fish or oil to planters' accounts, and he enforced this rule by attaching enough fish and oil in merchants' hands to pay wages.[17]

Throughout the early nineteenth century the Northern Circuit Court heard claims about pay deductions for alleged neglect and disputes caused by servants who resisted their masters' attempts to reduce their wage balances through extra credit charges for goods supplied on account. In 1826, for example, Robert Knox sold his servant, Timothy Mulcahy, £7 in tobacco, clothing, and tools during his time of service. Knox consequently had to pay Mulcahy only £17 out of his £24 wages at the end of the fishing season. Yet Knox would not pay this amount until Mulcahy sued. In court Knox bargained down his actual payment of wages by having the court deduct £6 for 17 gallons of rum, £1 neglect of duty, and 8 shillings for breaking a window. Mulcahy actually received only about £10 in wages. In most cases, however, the court ruled in favour of the servants.

The Northern Circuit Court sometimes tired of this battle of wages between masters and servants. When William Thistle withheld Michael Maratty's wages of £17 in 1831, Judge A.W. DesBarres decided for Maratty after deducting less than a pound for neglect of duty. The Judge accompanied this decision with the opinion that masters should be very cautious in withholding wages for such petty disputes in future.[18]

Servants, whether hired on fixed wages or shares, had to contend with the truck practices of their planters, particularly after 1824 when planters only had to reserve one quarter of their pay until the end of the season. This proved to be the case in the wage dispute between Thomas Shea and Timothy Crimin of Brigus in 1826. Shea had agreed to serve Crimin from 17 June to 31 October as a shareman for half the fish he caught, deducting 20 shillings for his berth plus a freight charge and a share of the cost of putting tree-bark rinds on board Crimin's schooner to build flakes at Labrador. Yet when he applied for his wages, Crimin gave Shea an account full of overcharges. Shea complained to the Northern Circuit Court that he did not authorize Crimin to take pork or flour on the servant's account with Pack, Gosse and Fryer, the supplying merchants. Crimin further charged too much for his berth, his rinds, and some tobacco and bait. The court allowed Shea 10 shillings for overcharges on bait but did not recognize any of the servant's other claims, leaving him with no wages due.[19]

Servants were not free from problems created by the widespread use of credit in the northeast-coast fishery, even when they signed what were apparently straightforward, fixed-wage agreements with planters. Masters paid wages only after they balanced their own accounts with merchants, which included charges for goods supplied to the former's servants. The 31 May 1827 indenture of Thomas Pyne to Michael Barry suggests that goods he took on credit ate up his wages when they came due at the season's end. Pyne agreed to serve as a fisherman or shoreman for one year in return for £18 wages, one pair of shoes, and 'one half the Balance of my ac/ct to be paid on the last day of October next and the other half the Spring following.' An attached account of Pyne with Barry shows that, by the end of his first year of service Pyne actually fell into debt as a result of serving Barry, owing £23 in liquor, ale, flour, tea, molasses, small amounts of cash, and damages for losing a skiff (£6). From the 1820s to the 1840s, other court cases indicate that fishing servants' wage agreements with planters were usually for credit during the fishing season, servants taking goods for the remainder of their wages if planters owed them any balance at the end of the fishing season.[20]

Planters like Thomas Deady of Harbour Grace enjoyed two advantages in using credit to pay servants' wages. First, supplying servants served the planter as a business in its own right. Deady hired William Fitzgibbon for £24 wages for the 1844 summer fishery, the balance of which was to be paid half in cash and half in goods at the season's end. Throughout the fishing season Fitzgibbon took on credit from Deady, leather, hemp, an oil jacket, cloth, blanketing, 100 pounds of pork, flour, molasses, tobacco, women's boots, and soap to the value of £12. At the end of the fishing season Deady only owed Fitzgibbon a balance of £12. Second, credit ensnared a servant, preventing him from entering the service of another if he could find better wages. William Donnelly sued planter Jeremiah Lee of Harbour Grace in 1851 because the latter hired Donnelly's servant, Joseph Gosse. Gosse had agreed to serve Donnelly on 3 May as a fisherman for £20, payable in cash and goods. By 26 May Gosse was working for Lee. Donnelly complained to the Northern Circuit Court that Lee hired Gosse knowing that the latter had already taken £5 in credit from Donnelly as his servant.[21] Credit reinforced the ties between servants and planters, and gave masters a means of minimizing (perhaps at a profit for themselves) the amount of wages they actually owed at the fishing season's end.

Servants did not distinguish between planters and supplying merchants, who received planters' fish and oil when they came into court to force payment of their wages. As in the 1833 case of Thomas Calvert versus his master, James Cuddihey, the court would order the supplying merchant (in this case, George Forward) to produce a full account of the planter's voyage, including a full list of wages due to servants. The court ordered both planter and merchant to pay if a servant could prove that the supplying merchant, as receiver of the voyage, owed wages. John Landergan sued his master, Carbonear planter Edmund Whiteway, and merchant William Bennett for his wages of £20. Landergan won £18, which Whiteway paid by an order drawn on Bennett.[22]

An 1832 case – James Brine's suit against the bankrupt estate of Harbour Grace merchant H.W. Danson – suggests that planters used merchant credit to pay wages to their servants. Brine demanded £55, his 1830 wages. The court awarded Brine £13, the balance of his wages, after deducting the supplies and diet he had on account from Danson. Servants seem to have been reluctant to accept payment for wages in cash if it meant that planters gave them a written order for cash to be advanced on credit by the merchants who supplied the planters. In 1832 James Conway sued Abraham and Joseph Bartlett for the balance of his wages: £16. The Bartletts told the court that they did not deny Conway the amount, and 'offered him an order on Mr Cozens for Cash But [Conway] would not accept of it.' Servants such as John Hunt, in 1832, seemed to want their wages paid directly by their masters. Hunt sued his master, Maurice Keene, for £13 wages, but would

[Handwritten marginal note, left margin: So far virtually whole chapter has dealt with credit & wages/shore for servants has → a pro/shore]

not accept an order by Keene drawn on 'his Merchant who is ready to pay the same when the fish is landed out of the Lady Ann.'[23]

Servants' reluctance to accept such planters' notes may have originated in uncertainty concerning just how the accounts of the planters and servants would balance on the merchants' books. Merchants did not pay wages to servants out of their own pockets, but simply on the demand of the planters' written orders. They waited until masters settled accounts to see how much planters would actually have to pay, and to see how much servants should actually get in cash after the servants' own debts were met for the season. Planter James Cuddihy of Carbonear gave his servant, John Healey, a note dated 23 October 1833 and addressed to merchant George Forward for £20 wages. Yet on 16 November John Healey was in the Northern Circuit Court trying to get the £9 wages due him. Cuddihy's voyage, as turned over to Forward, both the amount and the price received, was probably not sufficient to fulfil the total due. In 1833 Charles Kavanagh sued planter John Leary and the latter's merchant, William Bennett of Carbonear, for his £22 wages. Bennett produced an account of Leary's voyage to Labrador, which showed that Leary returned only £40 of fish against just over £150 wages he owed his crew (see table 5). The Northern Circuit Court ordered Bennett, as receiver of the voyage, to pay the servants' wages after deducting their accounts.[24] The court essentially ordered Bennett to pay the servants' wages in proportion to the amount of fish Leary actually turned in at season's end. Servants in this case used the courts to force reluctant merchants to pay wages out of their planters' fish.

When merchants faced suits due to planters' inability to process enough fish to pay servants' wages, they were not willing to assume the planters' obligations to their servants. William Bennett declared that he would only pay the servants of his planter, William Kehoe, to the extent of the fish and oil Kehoe actually delivered.[25] The Northern Circuit Court would balance the planter's account to determine the amount due the servants, as in the case of William Walsh's voyage in 1833 (see table 6). The entire catch of Walsh's voyage to Labrador could not cover his servants' entire wages.[26]

Servants' success in using the courts to enforce wage payments did not free them from bartering with merchants, particularly after the relaxing of the wage law under the circuit courts. Dennis Landergan and Michael Miles, two servants who worked for planter Joseph Puppy

TABLE 5
Statement of wage agreements and wages due to the crew of John Leary, 1833

Crew	Wages per agreement	Due from Bennett
Charles Kavanagh	£ 22.00.00 6/2	£05.17.03
Edward Cumming	21.00.00 6/2	05.11.09
James Hearn	22.00.00 6/2	05.17.03
Thomas Murphy	21.05.00 6/2	05.13.01
John Connelly	16.10.00 4/5	04.07.10
Catharine Maddock	06.00.00 1/7	01.11.11
James Breen	09.00.00 2/6	02.07.10
John Quin	16.10.00 4/5	04.07.10
Thomas Oats	10.00.00 2/8	02.13.04
Phillip Murphy	06.00.00 1/7	01.11.11
Total	£150.05.00 @5/ 3d	£40.00.00
Fish received by Mr. Bennett 80 qtls @ 10/} £40		

Source: PANL, GN5/3/B/19, HGCR, box 18, file 6, writ no. 282

at Labrador in 1832, complained that William Bennett would only give them fish as payment for their wages. In another case, Geoffrey Rielley agreed to serve Benjamin Leary for £21 wages as a splitter in 1833. At the end of the season Rielley had a £6 balance of wages due, which Leary could not pay in money. The court allowed Leary to pay the debt in fish. Rielley then had to bargain with a merchant over the price of his fish. In the same year merchant H.C. Watts paid wages in fish totalling £81 owed by one Hamilton to a crew of seven servants.[27]

There were ways planters might try to avoid paying their servants' wages, other than resorting to the complexities of credit. James Coyne, a Harbour Grace fishing servant, found in 1830 that his masters, planters Nathaniel and Thomas Davis, simply tried to avoid him when it came time to pay his wages. On their return home from a trip to Labrador, Coyne found that the two planters kept putting him off 'from day to day until their fish was disposed of and then they told his petitioner that it would be paid him as soon as possible.'[28] The manner in which other planters went out of their way either to avoid or run away from servants' wage demands testifies as to the latter's effectiveness in getting their payment through court action.[29]

Planters could also try to avoid paying servants' wages by using intimidation to get them to quit. For example, in 1832 Owen Fitzgerald

TABLE 6
Statement of William Walsh's voyage of fish and oil, and crew on wages the past
season, 1833

Fish & Oil received by Thos. Chancey & Co.		Men on Wages & 1 Woman	
172 3/4 qtls. mble. @	£86.07.06	David Laherty	£ 18.00.00
9 1/4 @	4.03.03	John Condon	18.00.00
66 gallons cod oil	5.03.02	Patrick Dunphy	18.00.00
	£95.13.11	Edward Nowlan	16.10.00
deduct freight	10.01.01	Edward Power	22.00.00
	£85.12.10	James Carberry	14.00.00
		Ellen Grady	07.00.00
			£113.10.00
A/C Shortfall for servants: £27.17.2			

Source: PANL, GN5/3/B/19, HGCR, box 20, file 1

complained that though he had worked for planter James Britt of Harbour Grace in the Labrador fishery for eighteen months on their return to Harbour Grace at season's end, Britt had so harshly criticized, belittled, and mistreated him that he considered leaving his service – except that he needed his wages. In a similar incident in 1854, servant George Mills complained that his master, Dennis Shea, tried to scare him off after their trip to Labrador with mistreatment and threats of beatings.[30]

However, resorting to relatively passive negotiation through the courts was not the only means servants used to combat planters' and merchants' use of truck against their wages. Servants would also respond violently to unfair treatment. Thomas Newell, the agent for Slade, Elson & Co. of Carbonear, complained in 1833 that, while engaged in settling the accounts of a number of the firm's planters, James Murphy had demanded his balance of wages out of 'his Turn.' Murphy, a servant of a planter named Luther who dealt with Slade, Elson & Co., responded by threatening to tear out Newell's throat. Murphy later apologized for his conduct, but Newell feared that Murphy's behaviour might encourage other servants to try to settle their accounts the same way. A similar case arose in 1837 when merchant Thomas Ridley complained that fisherman David Power, a shareman of one of his planters, made a drunken demand for money at his house. The merchant refused, leading Power to demand 'his account which com-

plainant told him he could have by applying at the Counting House.'
Power refused to leave and a fight ensued in which Power struck Ridley, who, with the assistance of James Gorman, threw Power out.[31]

Violence by servants did not always represent their desire to be paid
out of turn, nor did it always stem from too much drink. Servants
used force to express their frustration in trying to collect wages from
planters' accounts. In 1855 planter Henry Thomey of Harbour Grace
complained to Justice of the Peace R.J. Pinsent that he had met a former
servant, James Wilson, in the office of his supplying merchant, Ridley
and Sons of Harbour Grace. When Wilson demanded that his wages
be paid, Thomey told him that 'if he had any demands against me
– he knew what his remedy was,' implying that Wilson should go to
court. The servant instead resorted to a more personal remedy by
assaulting the planter, harassing him the next day at the wharf of his
supplying merchant.[32]

Servants sometimes joined together to force payment of their wages.
William Walsh's employees, fearing that their master would not pay,
took direct action against Walsh's supplying merchant, William Bennett, in 1833. Walsh abandoned his voyage to the servants and left
the dispute. His Labrador fish was green, and Bennett had to have
it cured before he knew how much quantity and quality he would
have to balance against the servants' wage claims. Three of Walsh's
servants came to the place where Walsh landed the fish and took what
they pleased before Bennett had a chance to settle accounts. When
Bennett tried to 'defend his property from such an illegal plunder he
was [told] by said Servants that they would kill him with stones if
he did not let them take away the fish.'[33]

The sometimes tumultuous disputes between planters and servants
in the northeast-coast fishery did not indicate a clear separation between the two groups along class lines. Paternalism linked masters
and men together in production, and they faced a common vulnerability
in the attempt to keep abreast of credit obligations, insolvency and
possibly starvation. Catastrophe could strike planters, as it did in the
case of the family of George Pynn of Musquito in 1828. Smallpox
struck the Pynns in May, and on 30 May the magistrates, fearful of
an epidemic, quarantined the family, using a guard to prevent any of
its members from leaving home. All George's plans for the fishery
lay in ruins: he could not send his schooner to Labrador, so his merchant, Bennett of Carbonear, withdrew Pynn's credit for the winter,
leaving his family with nothing to eat.[34]

A lesser but more frequent calamity to strike a planter was a call
from the sheriff or bailiff with a writ of attachment for some outstanding debt. The bailiff's call could be traumatic, joining planters
and servants together, as it did, in defence of the proceeds of their
labour. Charles Kennedy and Thomas Bartlett, planters and partners
in the fishery in 1833, reported that Bailiff Arnold Webber struck both
of them to the ground when he tried to seize their property in payment
for a debt.

The bailiff could be met with violence in turn. When Arnold Webber
and Martin Kelly attached the fish and oil of Michael Norcott in 1841,
Norcott threw rocks at them. The servants of planter Noah Perry assisted him in resisting an attachment on his boat by Bailiff Webber
'by throwing the said Arnold Webber out of the said Boat into the
sea.'[35]

Planters and servants at times faced common problems. When poor
catches, high prices for supplies, and low prices for fish coincided,
a planter's voyage could end in a merchant's taking him to court for
an account balance due. When the firm of Pack, Gosse and Fryer sued
him for £35 in 1829, Thomas Hedderson stated plainly that he could
not pay the debt due to a bad season 'and that it is with difficulty
he can get a little to support his wife eight children and an aged
mother.'[36]

After planter Daniel Meany ran a negative balance on his account
to Pack, Gosse and Fryer for at least a second year, the merchant firm
sued him for about £374 due in 1840. In 1839 Meaney owed the firm
over £226, but Pack, Gosse and Fryer risked an advance of £488 to
Meaney in supplies and wages for servants for a trip to Labrador.
Meaney could only return less than £341 in fish and oil to meet this
debt.[37] In 1844 Pack, Gosse and Fryer took another of its planters,
Timothy Morea, to court for over £51 left owing on his account after
a trip to the Labrador fishery. Morea took £187 in supplies on account
for the trip, but only returned roughly £136 in fish and credit against
the debt. An approximately £51 in debt does not seem like much, but
it appears that Morea tried to settle with his merchant without meeting
the fixed wages of his servants (Morea paid his sharemen). Morea
owed Patrick Moratty £23 wages, Richard Morea £30 (including £7
for his wife, Ann), and William Morea £22. Timothy Morea's total
debt thus equalled at least £126.[38]

Planters, like servants, faced ruin when they did not earn enough
to pay their accounts, and they sometimes blamed supplying merchants'

credit manipulations for their failure. Planter Thomas Powell of Carbonear complained in 1827 that he had carried on a prosperous fishery for nine years, supplying Pack, Gosse and Fryer with seals, fish, and oil in return for credit. Powell fared well until 1825 when the fishery failed. The planter ran a negative balance of £62, so Robert Pack refused him further credit. Pack attached Powell's share in a schooner held in partnership with the firm, preventing Powell from going to the spring seal fishery and ending the planter's last hope of meeting his debts. A sheriff's sale disposed of Powell's share to Pack for £21, although the Harbour Grace Insurance Society had valued the schooner at over £200. Another sheriff's sale sold Powell's room and stage for £5 (although valued at £25), his fishing room and plantation at Carbonear, and a stable, barn and dwelling to Mr Pack to satisfy the remainder of the debt. Powell left no doubt he felt that merchants, not a poor season, had brought about his downfall:

Your ... [memorialist] therefore feels himself greatly aggrieved, in being thus pressed for a debt for goods charged at exorbitant rates, far higher than Mr. Pack himself sold them for to others of his dealers, and which, if they had been reasonably priced, would have left no balance whatever due to Mr. Pack; in Mr. Pack's preventing the schooner from earning money in the Spring of 1826, between the test of the writ and its return; the sale of the share of the vessel for £21, which was even by Mr. Pack's own valuation worth £70; the sacrifice of his other property to satisfy a rapacious and merciless creditor.[39]

Powell accused his merchants of using truck to ruin him. William Morey likewise suspected his supplying merchants, Rogerson & Cowan, of keeping strange accounts when they attached his plantation for a debt of £97 in the fall of 1826. Fisherman John Mason found that his supplier, John Hackett, had actually overcharged him in 1829. Hackett claimed that Mason took £41 in goods against which he returned £37 for a balance due of £4. The Northern Circuit Court investigated the account, found the overcharge, and reduced the debt to £2.[40]

Planters, like their servants, not only felt ill-used by truck at times, but could also occasionally suffer similar physical mistreatment at their merchants' hands when they disputed their accounts. Planter Jacob Nicholas of Harbour Grace complained of an assault by merchant Henry T. Moore in 1854. He stated that, when Moore gave him his account containing objectionable overcharges, Moore 'took me by the collar and dragged me to the door – and kicked me violently with his foot.' Moore replied by stating he took this action after Nicholas used foul language in response to his reading of the account.[41]

Planters could try to use merchant credit to their own advantage in dealing with their servants, but in turn could expect little leniency from merchants' accounting. The petition of Michael Keefe gives some idea of the hazards faced by planters in relying on merchant credit in large-scale operations. Keefe had built up an extensive fishery, dealing with a number of merchants over the years and employing many servants. But between 1832 and 1833 he fell into debt to his supplying merchant, J.C. Nuttall. The merchant sued Keefe for the debt, refusing to honour a verbal agreement that the planter would pay off the debt at £8 per year. Nuttall had Keefe jailed for the debt on 12 December 1833. Nuttall defended his action during Keefe's insolvency hearing by charging that the planter had begun to transfer his property to his sons in an attempt to elude his creditors, and would not sign a bond for the £8 per-year agreement. Nuttall suspected that Keefe was trying to evade his credit responsibilities to Nuttall as his supplying merchant, beginning in 1832 when he sold fish to the Slades at Battle Harbour, Labrador, for lower-priced provisions. Nuttall made no apologies for no longer being willing to trust Keefe, and in turn asked the court to secure his property. The court declared Keefe insolvent.[42]

While debt-ridden planters felt that merchants' manipulation of prices caused their problems, there exists little hard evidence that this was the case. We have only generalizations about truck, such as that of an 1829 observer who attributed most of the colony's economic problems to merchants' ability to place 'an arbitrary nominal value on the fish at the end of the season, and the supplies are charged more with reference to that supposed value, than to their actual cost of importation; consequently the Merchants rely for their profits almost entirely, on the imports, for generally speaking, if fish realizes its nominal cost, and charge of freight in the market, it is accorded a successful speculation.'[43] Through the 1840s and 1850s much public debate took place within the Conception Bay press about class relations in the fishery, which suggests that truck had some advantages for planters and servants. Some pseudonymous newspaper correspondents, such as one who called himself 'A Voice From The North Shore,' felt that the ongoing depression in the fish trade could not support free market relationships between merchants and their dealers. Merchants, by not undercutting each others' prices in an attempt to gain more customers, cultivated a paternalistic bond with planters, to avoid losing the latter's fish in a ruinous competition with each other. Supplying merchants could count on planters to return fish and oil on their accounts year after year, if the former provided credit during bad years. A good supplying merchant was one who acted as 'a father towards his planters.'[44]

Not everyone accepted this image of merchants as benevolent paternalists. Another correspondent described the fishermen of Conception Bay as worse off than the 'slaves and serfs of Russia,' subject to their supplying merchants' humour when they begged for credit at the end of the fishing season.[45] But aside from such anecdotal assertions, there is little to substantiate that any merchant in the Newfoundland fishery of the first half of the nineteenth century pursued a debt-led strategy to secure fish supplies, or used unequal exchange with planters by fixing fish prices relative to the prices for supplies given out on credit for the fishing voyage. All that can be said for sure is that truck consisted of merchants exchanging capital goods, supplies and provisions in return for planters' fish and oil. The governors' returns state that, from 1815 to 1825, merchants in Conception Bay set fish prices at current international market prices. At Trinity, merchants gave either a 'general price,' the 'Market price,' or the 'Newfoundland price.' The returns generally reported that the merchants at Bonavista and Fogo usually gave the market price for fish, just as they did in Conception Bay.[46]

Planters could even be drawn into a certain identification with merchants when they seized on the opportunities suggested by trade with their own servants, withdrew from the risks of production, and began to trade on their own account. Charles McCarthy, a prominent merchant in Conception Bay in the 1840s, started out as a planter at Crocker's Cove. McCarthy seized on the opportunities posed by the insolvency of Bristol and Harbour Grace merchant H.W. Danson, in 1831, by purchasing Danson's brig *Curlew*, and using it to trade on his own account. In 1839 planter Richard Taylor took a more restricted step by hiring his schooner to merchants such as Chancey & Co., to carry supplies to and fish from the planters on the Labrador. By 1847 George Udell of Carbonear was using his schooner to carry other planters to the fishery at Labrador.[47]

The risks of mercantile activity reinforced the perceptual bonds which linked planters and merchants together in paternalism. H.W. Danson became bankrupt as a result of his investments in England.

In the financial panic there in 1831, his English assets lost much of their value. A representative of Danson's English estate asked his Newfoundland trustees to sell off his assets. James Prendergast replied for the Newfoundland trustees that Danson's Newfoundland assets, consisting mostly of overvalued mortgages on planters' operations, were overrated, and would not pay 2.5 shillings for every pound owed. Danson held mortgages valued from £90 to £1,200 and totalling almost £6,000.[48] Merchant William Innott became insolvent in 1833 when he could not meet the demands of his own creditors. While Innott owed forty-four creditors £1,658, more than 130 people owed him £3,543. Innott could not find a way to make these outstanding accounts satisfy his own creditors.[49]

Fish merchants found themselves caught in the same trap as fish producers. 'D' wrote to the *Weekly Herald* in 1852 that the past thirty-five years had seen the slow attrition of merchants in the fish trade. The old West Country firms which could not survive the credit demands of the fishery withdrew and left the trade to a more restricted Newfoundland-based operation. A supplying merchant found little satisfaction in the fish trade because 'his hard money goes to pay duties and other expenses of the trade. Short catches, ruinous markets, and desperate debts keep him perpetually upon the rack, and his situation is the unenviable one of being almoner to his dealers.'[50]

Other correspondents, such as 'Libra,' agreed with 'D' that merchants faced adverse trading conditions in the Newfoundland fishery. He did point out, however, that truck, the very means by which merchants tried to survive, was a form of exploitation of planters and fishermen. Merchants charged mark-ups on prices to profit from trade with fishermen in an economy that could only support merchants' profits or producers' profits but not both, because of its vulnerability to cyclical market or catch failures.[51] 'Alpha,' in the following months, defended merchants by asking what alternative to merchant capital existed in the Newfoundland fishery, and if the merchants' critics accepted the principle of private enterprise in the fishery. Merchants were not in the business to carry planters and fishermen on their backs, nor were they there to develop other sectors of the Newfoundland economy, unless some good reason existed to do so, and without a local agricultural base, none did so.[52]

The paternalism inherent in merchant supply may well have blunted the edges of the class struggle within the fishery. Planters and servants could rally to the support of merchants in their communities as easily as challenge them through the courts or with violence. But, as in the case of the Leamon affair, such support could be very ambivalent. When men broke into Leamon's store and stole much of the store's clothing and equipment, the sheriff of Harbour Grace, G.C. Gaden, came to Brigus to investigate the affair. His deputy sheriff, Johnston Burrows, after talking to Leamon's son, Robert, when he first served the attachment of Bowrings, heard from the latter the suspicion that lay in the officials' minds: Brigus people would not allow their merchant to be closed down by Bowrings, a St John's firm. If, on the other hand, the sheriff could convince them that the attached goods still belonged to Leamon, they would not try to steal them. Burrows took the precaution of removing provisions to another store so the people of Brigus might be assured that they would not lose their winter supplies. Despite Robert Leamon's promise that the people would not take any other attached goods, the sheriff suspected that they had done so.[53]

Sheriff Gaden interviewed two groups of people who lived in the Brigus area: people who worked in the fishery and those who did not. The latter, sure that Brigus fishing people had committed the robbery, displayed a certain dislike of them in the process. The Brigus schoolmaster, James Power, for example, said, 'I never suspected that it was strangers that committed the robbery. I thought it was the people of Brigus or the vicinity that committed the robbery.' Power disclaimed any suspicion about whether or not friends or enemies of Leamon did the deed. Accountant Joseph Cozens also had no suspicions in this last regard, and took pains to make clear to the sheriff that, 'I never heard Robert Leamon say that it would be a good thing to secure the goods so that they may not fall into the hands of his father's creditors.'[54]

Robert Leamon denied leading local people in a conspiracy to keep his father's goods out of the hands of the Bowrings. Fisherman Thomas Stephens Jr likewise denied participating in any robbery, or knowing anything about it. Many other fishermen and planters, either dealers of Leamon's or not, denied any knowledge of the affair. One fisherman, John Clarke Jr, denied that the Brigus people had anything to do with the robbery, stating that strangers to Brigus must have committed it.[55]

Three fishermen told Sheriff Gaden that they thought local people did it. Leamon owed these fishermen money, and they probably saw their chances of recovering any part of their debt diminish with the theft. John Cole, owed between £100 and £150 by Leamon, said, 'I heard some person say that perhaps it might be some of Mr Leamon's people that took it. I imagined from that expression that they meant it was taken for his benefit.'[56] Nathan Clarke, who Leamon owed about £35 for seals, also felt sure that Brigus people committed the robbery.[57] Fisherman Thomas Stevens, after telling the sheriff that Leamon could not repay his debt, stated 'I think it was persons belonging to Brigus or handy about the neighbourhood that committed the robbery.'[58] The statement of fisherman Michael Merrigan may have confirmed the feeling of law officials that Brigus people took what they saw as their goods to keep them from falling into the hands of a St John's merchant. Merrigan, living at the home of John Power, stated that his fellow lodger, John Lundregan, implied to him that local people took the goods. Lundregan told Merrigan, who did not participate in the affair, 'Your harm is done you are too late now you will get no salvage out of it now.'[59]

The loyalties to Leamon, which officials must have felt existed on the part of some fishing people, can be seen in the nature of some of the answers Sheriff Gaden received to his questions. John Way Jr, who worked as a shipped servant to Leamon, denied that he took part in the robbery. He said, 'the devil secure the creditors or that I was not sorry for it.'[60] In a similar statement, fisherman John Sullivan responded to questioning by saying, 'I have never heard any person state either directly or indirectly that the property was taken by friends of Mr Leamon's. I never heard any person say that they were rejoiced that it was taken or words to that effect. I never heard any person say that they were sorry ... I never heard any person say that they were glad that Bowring had lost the property, or Mr Leamons creditors.'[61] Eleanor Dunphy, wife of fisherman John Dunphy, and a Brigus resident for thirty-three years, argued to the sheriff that she knew nothing of the incident, was sorry it happened, and never said 'that I was glad of it and hoped that the goods nor the people that stole them would never be found out.' John Dunphy confirmed his wife's statement.[62]

The investigation into the Leamon affair reveals the equivocal nature of paternalism. A number of people, for the most part judicial officials, other people who did not fish, and fisherman-creditors of Leamon, were certain that local people took the goods that were under attachment in Leamon's store. No one, however, was certain whether or not Brigus fishing people committed the robbery to support the Leamons or simply to obtain free supplies for themselves. It may be that some people in Brigus committed the robbery out of loyalty to Leamon, or they might have taken the goods with no intention of giving anything to Leamon. What is clear, however, is that the people of Brigus saw Leamon's mercantile activities as an essential part of their community, one that could not be allowed to fall into the hands of a St John's firm with no commitment to Brigus. A mutual, although unequal, accommodation of each others' needs forced fishing people and fish merchants together. In particular, as is clear from Leamon's son's testimony, the people of Brigus saw Leamon's credit as a much-needed source of the provisions that allowed them to live. Robert Leamon felt sure that Brigus residents would deny Bowring the right to attach the foodstuffs of their merchant.

Credit permeated every layer of class relationships among those involved in the making and trading of salt cod on the northeast coast. Merchants were not the only ones to deal in truck with fishermen. Planters used truck credit practices in the accounts they kept with their servants. Just as planters relied on their supplying merchants, servants relied on their planters – either on direct account or indirectly through their merchant – to provide on credit the provisions and equipment which made life possible in their fishing communities.

There is little doubt that servants and planters who lost the faith of their merchants and ended up without credit, insolvent, or possibly in jail, felt unfairly treated. Merchants exploited servants and planters in that they wanted to profit from the credit they extended to both. The paternalist chain of credit which linked merchants, planters, and servants at all levels of society in mutual necessity was a lasting one. This is not to say that class struggle did not exist in the fishery. Male servants resorted to the courts and to violence to force planters and merchants to honour their obligations in truck; planters, at times, resorted to similar tactics in dealing with merchants. Nevertheless, credit offered some planters opportunities in petty trade that it did not offer servants. In all cases truck was a negotiation between planters, servants, and merchants, which in the absence of any alternative development, persisted throughout the first half of the nineteenth century. The prospect of violent confrontation, combined with the wage and lien system, may have encouraged many planters to retreat from hiring male labour.

PART FOUR: THE CHIMERA

7

Agriculture and Government Relief

'You will ruin the Colony by such extensive employment' of Conception Bay residents on relief work, Colonial Secretary James Crowdy warned Justice of the Peace Robert Pinsent of Harbour Grace in 1847.[1] Despite more than twenty years of unrestrained land alienation and concerted government support of agricultural improvement, local magistrates throughout Newfoundland found fishing people still dependent on government relief for survival. Yet the cost of such assistance so strained the colony's revenues that officials were desperate to believe agricultural development could succeed. Colonial authorities clung to such unrealistic optimism to avoid the despondency of accepting that, without some fundamental reorganization of the fishing industry, government might have to go on supporting Newfoundlanders without foreseeable end. While land and climate frustrated government policies, little could be done about these. But their need to establish some efficacy led officials to grasp for other reasons for failure, and this made them susceptible to one of the messages of Newfoundland reformers: that the colony had good resources, but awaited only some encouragement after years of restraint under the fetters of merchant capitalism. The Newfoundland governors came to believe in this agrarian myth, even as their encouragement of agriculture failed to alleviate the problems families faced in a depressed fishery.

St John's reformers such as Carson and Morris had been using the agriculture issue to agitate for representative colonial institutions between 1818 and 1824. They got instead, in 1825, a civilian governor who was keenly interested in the colony's internal improvement, but no legislature in which reformers could play a direct role in government. Governor Cochrane recognized that fishing people faced pro-

visions crises because of merchants' credit restrictions. Acting on the
advice of George Coster, the Anglican clergyman at Bonavista, Cochrane at first arranged to have food and seed potatoes distributed to
people by local merchants. While government had paid for these goods,
relief recipients believed they had gotten them on credit, which they
would have to repay when possible. Rather than confront the problem
of credit restriction in the fishery directly, Cochrane chose a short-
term solution essentially comprising a government subsidy of the merchant supply system. For the long-term the governor turned to the
promotion of farming as an alternate source of food for residents.
The governor continued to give out relief, but required in return that
recipients build roads to open up interior land. He hoped, once residents had cleared the land and begun to farm, people of means might
invest in estates, become a landed gentry, and provide a suitable social
basis to warrant a legislature for Newfoundland. Cochrane's policies
did not work; no matter how many roads were cleared or how much
seed was distributed, farming did not thrive. The governor reconciled
himself to the colony's relief burden by 1827, and opposed representative institutions as an additional obligation Newfoundland could not
afford.[2]

Cochrane's disappointments played into reformers' hands. Rather
than admit that Newfoundland was simply not suitable for agricultural
colonization, because its resources were comparatively deficient, Patrick Morris claimed that the colony was not matching the pace of
internal improvement set by Nova Scotia, New Brunswick, or the Canadas because the latter three had legislatures and Newfoundland did
not. Morris and other petty merchants had hoped the governor would
grant them large tracts of land so that they might raise crops to sell
to their fishing clients. Morris felt that such estates were justly due
him and his confrères, as Newfoundland's local elite; furthermore, that
they deserved a legislature in which they could aggrandize themselves,
as did colonial gentries elsewhere, and oppose the governor when he
denied them.[3]

The appropriate response to reformers' demands was really quite
simple, thought Cochrane: let a local trader like Morris lease land,
farm it profitably, and pay his dues to government if he could. If enough
landholders succeeded, proved they could hire labourers and sublet
to tenants, and made their remittances, then let the Colonial Office
grant Newfoundland a legislature. The governor called the reformers'
bluff by suggesting that they develop the land themselves, and prove

that the island could support a gentry and representative government. Until they did so, there was no reason for Cochrane not to accept the advice of Chief Justice Tucker, as president of the executive council, that dependence on the fishery made Newfoundland little more than a 'common manufactory' which needed sound management but not any measure of self-government.[4]

Morris, Carson, and other members of the St John's bourgeoisie almost immediately received the land grants they desired. The Newfoundland government, in 1831, allowed leases at the low rate of nine pence per acre, but the reformers began to complain that they could not earn enough from their farms to pay even this pittance.[5] Instead of admitting to their own failures, the reformers chose to criticize government for not doing more to encourage them by promoting the full-scale agricultural colonization of the land. Governor Cochrane thought that such an effort would be folly. Instead, the furore developing over Crown rents was politically motivated by reformers wishing to gain support by spreading false rumours about government intentions. Cochrane pointed out that government only tried to collect rents from the small bit of commercially oriented land around St John's that was owned by people like Morris and Carson. Land which had always been cultivated by fishing families in the colony's outports was not subjected to severe exactions by government. Charging that reformers were merely riding the wave of agitation over Crown rents that was rising in other parts of British North America, Cochrane was amazed that anyone might take reform assertions about Newfoundland's agricultural potential seriously if land already farmed could not bear a small rent.[6]

The reformers decided to remove their struggle overseas to a new arena, attracting the attention of British parliamentarians and Colonial Office personnel, who were tiring of Newfoundland's dependence on grants from the imperial treasury to pay its relief bill. In a letter to the Royal College of Physicians in 1831, William Carson repeated the claim that Newfoundland suffered under the burden of a mercantile-government conspiracy to prohibit settlement in favour of a migratory fishery and merchant profit. The result was that Newfoundland did not have the refined elements required of gentry society: roads, agriculture, and well-developed educational institutions. Carson condemned West Country merchants as still being opposed to colonial self-government and settlement, but applauded the more 'paternal' British authorities for gradually giving the colony its due. Encourage-

ment already given to agriculture by grants and road-building had led
to great improvement, although this still largely took place around St
John's. Newfoundland needed more. The country supported a vigorous
population of true Britons who deserved agricultural improvement as
their right, just as colonists did in the other British North American
colonies.[7]

The assumption underlying the seductive appeal to the Colonial Office for reform was that Newfoundland had the same resources as
any other British North American possession. Although Cochrane sent
an accompanying caution with Carson's letter, in which he warned
that he knew of no other place in the empire with as high a proportion
of 'sterile and useless ground' as Newfoundland, Carson's message
had a powerful attraction for a Colonial Office tired of dealing with
continuous crisis in the Newfoundland fishery. The reformers claimed
that depression in the fishery had released an abundance of labourers
ready to work cheaply on farms, awaiting only the Crown's alienation
of large tracts of land without rents or charges.[8] Carson held out the
hope of a gentry who would employ those who could not be supported
by the fishery. Government relief payments consequently would abate,
but only if the recalcitrant governor would relax his efforts to see
this aspiring gentry pay something for their land. It was essential that
a local legislature start administering land and internal development
policies properly.

Cochrane remained steadfast in his opposition to the idea that Newfoundland should have representative government, pointing out that
his constant relief of outport people proved that the reformers' vision
of a gentry with a base in commercial agriculture was but a pipe dream.
He scorned the agricultural claims of people like Morris, noting that
their own cultivation at St John's had been no more than an effort
to claim arable land as their own and then to sell it at a profit to
neighbouring fishing families. The reformers' misrepresentation of
Newfoundland's resources completed Cochrane's disillusionment. He
reported that, although he had been enthusiastic about Newfoundland's agricultural potential early on, his increasing familiarity with
the island had convinced him that it was a barren place, fit for little
but fishing.[9]

The best thing government could do for Newfoundland, Cochrane
argued, was to use relief to allow people to subsist until such time
as the fishery might recover. While it was necessary to make people
work on roads for such assistance, and to have them work their gar-

dens, this stemmed more from the governor's belief that free aid would make people indolent, rather than from any faith in the potential bounty of the soil. Chief Justice Tucker supported Cochrane, but went further by arguing that any great encouragement of farming by government would fail to do anything but drain the revenue and countenance an unrealistic optimism that could well lead to Newfoundland's trying to live beyond its means.[10] The sentiments of Cochrane and Tucker did not stop the reform movement from slowly gaining momentum, one that would persuade succeeding government officials to accept the notion that Newfoundland had agricultural resources that some cabal of non-native merchants and officials must have purposefully left underdeveloped.

Merchants in the most densely settled parts of Conception Bay began to support reform views. In 1831 Thomas Ridley of Harbour Grace and Robert Pack of Carbonear organized local support for the St John's reformers' demands for representative government. At a 4 October meeting Pack made the reason for his support clear by publicly thanking Carson for his twenty years of continuing support of agriculture as a supplement to the fishery. These merchants were colonial residents who depended on family agriculture to minimize the need for credit in the fishery. Ridley and Pack, in addition, probably looked to representative government as a way of lessening St John's influence in colonial administration.[11]

Believing reform promised that a local legislature could provide the funds for the colony's internal improvements, the British government saw a way to absolve itself of responsibility for Newfoundland's revenues.[12] Reformers' demands fell on increasingly more sympathetic ears at the Colonial Office in London. Colonial reformers Edward Wakefield, Joseph Hume, Charles Buller, and Sir William Molesworth argued that the lack of self-government in the colonies exposed British subjects to the despotic and arbitrary rule of colonial governors, and incurred unnecessary expenses for the British government. The Colonial Office bowed to the weight of such criticism, accepting a more informal imperial structure in which self-governing and self-taxing colonies were to pay their own way as much as possible.[13] The Newfoundland fishery was no longer indispensable to an empire dominated by industrial capitalism and its free-trade ideology. Great Britain's rulers had accepted Adam Smith's fiat: colonies which could not or would not aid the empire's support must be cut adrift to look after their own civil and military needs.[14]

The British government granted representative government to New-foundland in 1832 partially because its desire to lessen its responsibility for the troublesome colony made it susceptible to a belief in the developmental potential of Newfoundland agriculture. But even with representative government, Cochrane had to spend much time pleading with the British government to finance relief of fishing families. The winter of 1831–2 had been unusually severe and long, again forcing the government to issue more seed potatoes to avert famine. In Conception Bay, mobs had begun to loot merchant stores for bread and other foodstuffs. Such conditions persisted through 1834, leading Cochrane to ask the British government to grant even more funds to the Newfoundland government for relief. The governor constantly pointed out that Newfoundland revenue rested almost solely on customs revenue, which fell with the fishery's decline. Reformers' demands that no rents be taken in exchange for land alienation meant that government would lose its only other possible local source of revenue.[15]

Cochrane received many expressions of displeasure from Colonial Office officials, who had believed reformers' promises about the ability of a Newfoundland legislature to minimize relief through the encouragement of agriculture. Colonial Secretary E.G. Stanley informed Cochrane that reform demands for a legislature had been granted because reformers had promised that local economic development would lessen, not increase, expenditures on relief as well as pleas for more financial aid. Now the colony wanted more British assistance at a time when the empire was cutting its aid to the other British North American colonies. The message was clear: the Newfoundland government could no longer look to the Colonial Office to foot the bill for its relief problems.[16]

The experience of northeast-coast fishing people does not provide much evidence that the Newfoundland government would have been able to support its own population if it had had to rely on official encouragement of local agriculture. It must have been quite worrisome for British and Newfoundland officials to again observe crowd actions in response to the inadequacy of merchant credit for winter supply and government relief. The agent of the merchant firm James Mac-Braire & Company at King's Cove, Bonavista Bay, reported that a crowd from King's Cove, Keels, Tittle Cove, and Stock Cove showed up on 7 June 1832 – with the ice not yet free to allow a ship in with relief supplies – threatening to break open the doors to their stores unless given bread. The agent gave out food to avoid violence. In 1833 a similar group forced the stores at Catalina. To keep the crowd of 1816–17 fresh in the mind of the government, Justice of the Peace Thomas Danson wrote almost annual requests for compensation for his part in relieving from his stores the earlier mob at Harbour Grace.[17] The government witnessed history repeating itself in the crowd actions of 1832 and 1833.

But rather than accept that Newfoundland simply did not have the resource base to permit the success of farming, the Colonial Office chose to see the problem as a failing of the governor. It recalled Cochrane in 1834, in part because of his intensifying debate with the reformers (by then popularly known as Liberals), and possibly also because he could not lead the colony to independence from British financial support. Yet the Legislative Council, dominated by conservatives, again petitioned the Crown for money to supplement the revenue. What was worse, the council held out no hope of future improvement, telling the Colonial Office that agricultural diversification would never be successful and demanding a bounty on Newfoundland fish.[18]

In 1837 both the cod and seal fisheries failed, leading people to look again to the new House of Assembly for winter relief. The Newfoundland legislature had taken on a governmental role similar to that established by Cochrane, namely using road-work as a public relief measure and a means to give fishing families access to land. The inhabitants of places like Carbonear hoped that roads would improve winter access to their towns, so that families on the north shore of Conception Bay and the south side of Trinity Bay would have easier access to town merchants' stores in case of provision shortfalls. After almost fifty years of unofficial and official encouragement of cultivation, some Newfoundlanders still thought successful agriculture lay just around the corner, awaiting only some new encouragement from government. 'Public Opinion' hoped that a House of Assembly-sponsored road bill (one which had to be approved by a hostile Legislative Council) would create a new age of agricultural prosperity in Newfoundland. The editor of the *Sentinel* at Carbonear was not so optimistic, noting that government had little revenue to spend on roads, and would be better off improving the fishery by giving bounties to Newfoundland fishermen.[19]

The fruits of such 'encouragement' to agriculture were harvested by fishing families in the winter of 1838–9, when men from Trinity

Bay walked across the barrens to Harbour Grace to obtain relief when their provisions ran out, reporting that their families were starving.[20] In 1839 a petition from 142 residents of the north shore of Trinity Bay, begging for help to prevent famine, forced the assembly to recommend relief. Information from the justice of the peace at Port de Grave suggested that similar conditions prevailed in Conception Bay. The Assembly resolved itself into a 'committee on seed potatoes,' so that it might press the governor to purchase 325 pounds of potatoes to distribute throughout Trinity and Conception Bays in accordance with the old Cochrane-Coster plan.[21]

Despite reports of a good fishery in the summer of 1839, the winter of 1839-40 proved to be another season of food shortages. Neither the fishery nor local cultivation seemed able to do much more than allow people to limp through winter. Some debate began to develop about the wisdom of government continuing to support agricultural improvement because it had so far proved futile. One correspondent of the *Sentinel* suggested that improved conditions for Newfoundlanders could only come from a stronger fishery. While this 'Friend to Enterprise' confined his thoughts concerning the manner in which the fishery could be bettered to the need for imperial bounties, he had at least taken a step forward in addressing problems in the only industry in which Newfoundland had some comparative advantage. Unfortunately, changing the fishery would take a lot of time and political will, while the needs of residents were immediate. Other correspondents argued that even if road-work proved of little avail in encouraging agriculture, it at least provided people with short-term relief.[22]

Popular hopes for improved economic conditions consequently focused on the assembly's internal improvement schemes. The *Sentinel* condemned the manner in which the Executive Council allowed funds only for road-works in the St John's area in 1840. Through 1843 Robert Pack led a local effort to complete roads to Trinity Bay and along Conception Bay's north shore, hoping that the roads would create better economic conditions for the poor by providing them with access to even more land for clearing. Such encouragement to agriculture did seem to produce some prosperity in Conception Bay that year. In addition, reports suggested that the cod fishery would yield a catch good enough to allow fishermen to clear their accounts and perhaps establish a little credit for the winter. Crops in the bay appeared to be thriving by late summer, and the *Sentinel* noted that some people found they could even sell their early potatoes in Carbonear.[23]

Cochrane's successors, particularly Governors Harvey and LeMarchant, tried to win support from the reform-dominated legislature by continuing to encourage opening access to waste lands through road-building. Local enthusiasm for such programs among the St John's bourgeoisie led Governor Harvey in 1841 to remind them that agriculture could only succeed if it remained a supplement to the fishery. Despite this caution agricultural planning continued, and in 1841 a number of St John's residents formed the Newfoundland Agricultural Society, with a mandate to distribute seed potatoes, grain, and grass seed, as well as agricultural information throughout the colony.[24]

Government officials began to think that perhaps the reformers were right about Newfoundland's supposedly great agricultural potential. Even the *Public Ledger*, usually hostile to the reformers, agreed with them that Newfoundlanders should support the newly formed Agricultural Society. Its editor stated that Newfoundlanders had to try to succeed at agriculture, although the colony's soil and climate made success extremely unlikely. Unless they did so, however, Newfoundlanders would have to accept dependence on the fishery alone.[25] They did not have much choice in this grasping at straws.

Faced with merchants' restriction of credit, a series of crop failures, and constant demands for relief, the government again turned to agriculture as the panacea for the colony's troubles. In 1843, after the Assembly received more petitions for relief from Fogo, Tilting Harbour, Moreton's Harbour, Trinity and Bonavista, it formed a select committee to tackle the details. At the same time, Governor Harvey decided that the government should give all possible assistance to the Agriculture Society in its attempts to foster the cultivation of grains, turnips, and better potatoes. In Harvey's estimation, the salvation of Newfoundland lay in discovering the potential of the 'extensive Prairies of the interior of the Island for Cultivation and Settlement.'[26]

Popular disillusionment with government roads and cultivation plans could quickly appear, and Newfoundland's climate dimmed such optimism in 1844 when early frosts severely damaged potato crops.[27] After facing a hard winter in 1844–5, correspondents began to ridicule the government's attempts to introduce sheep husbandry to the outports. No one could understand how government could expect the land to support enough sheep to provide wool to clothe the entire outport population when the land could not support people in the fishery.[28]

The year 1845 proved ominous for the advocates of local agriculture in Newfoundland because it marked the arrival of the first serious potato blight in the island. The blight destroyed most of the northeast-coast crop, forcing the people to eat the seed potatoes they would ordinarily have used for the next year's planting. The Harbour Grace *Weekly Herald* advised its readers to take note of the Newfoundland Agricultural Society's recommendation that people try not to eat the seed potatoes issued as relief. In Newfoundland, by the mid-nineteenth century, subsistence standards ebbed and flowed almost as much with the success or failure of the potato as they did with the fishery.[29]

The spring of 1846 saw another provisions crisis in Conception Bay as merchants found that a poor fishery did not allow them to give out goods on credit. The summer's cod fishery failed in Trinity Bay, Bonavista Bay, and northward, while those in Conception Bay and Fogo proved mediocre. The poor returns of the seal fishery left many families without the means to pay for provisions. The winter of 1846-7 proved to be a disaster when extremely cold temperatures froze and spoiled an already small potato crop. In Conception Bay this meant 'that the great bulk of the population ... are totally destitute of the necessaries of life.' By spring, people in Conception Bay looked forward to the seal fishery employing able-bodied men, leaving only women, children, and the aged to depend on government relief. But again the seal fishery failed, forcing many people in Conception and Trinity Bays to eat seed potatoes to survive, and turn to government for the minimum relief of new seed potatoes.[30]

Government, unwilling to spend much on seed potatoes, provided relief only reluctantly. The *Weekly Herald* advised fishermen that the only way they could maintain any credit with merchants was to 'learn to be industrious and economical and honest and, if possible, independent' by taking special care in catching fish and cultivating potatoes. Despite such advice families in Trinity and Conception Bays continued to experience food shortages, and had to eat their seed potatoes before the summer of 1847 began.[31] The year 1847 was one of particular crisis for the people of the northeast coast. Charles Cozens, chairperson of the local relief commissioners in Brigus at Conception Bay, told the governor how everything had gone wrong that year. The winter of 1846-7 saw a fire in St John's which destroyed many mercantile establishments and made merchants unwilling to extend credit to the outports. Gales had destroyed much property, the seal fishery had failed, and – the last straw – potato blight swept the coast. Cozens feared that, because government had not made adequate provisions for relief, Conception Bay families might starve before the winter's end.[32]

Again the government decided to 'relieve' people by applying the Cochrane-Coster seed-potato plan. Unfortunately, the governor found that the usual source of seed potatoes – the small number of commercial farms which had grown up around St John's – had dried up, and potato blight hit St John's just as it had other parts of the colony. The Agricultural Society proposed using grain seed as a substitute for seed potatoes, and again admonished people not to eat what seed potatoes they had in hope of preserving them for planting the next year's crop.[33] The failure of local cultivation and the paucity of government relief led the editor of the Harbour Grace Weekly Herald to give fishing people some hard advice. If people could not raise their food and merchants would not advance any credit, then let them eat shit. The Weekly Herald's editor made his recommendation in language designed not to offend the delicate sensibilities of his readers, but this makes the alternative he offered to fishing people no less appalling: 'many a poor family during the course of the past spring was obliged to put up with ... a morsel of stale seal or a rusty herring, who, had they been more provident over what is considered by too many in this country as the refuse of the voyage; viz: the nutritious heads of the cod fish, the tongues and other internals, would in all probability have felt but little of the distress which they were forced to experience.'[34] While perhaps nutritious, such effluvium had usually been left to seagulls and other scavengers, or it had been piled outdoors for use in gardens; as Dr Carson had reminded people five years earlier, the heads and guts made 'large and fertile heaps of manure.'[35] Fishing people did not make a habit of eating what they saw as fish offal.

But people did adapt to newer, more repugnant ways of surviving as the crisis on the northeast coast continued unabated. Yet another poor fishery and a small potato crop due to the spring shortage of seed potatoes led the paper to forecast another winter of distress in the early fall of 1847. Potato blight hit Conception Bay hard, leading many people to seek relief in St John's.[36] John Soaper, an itinerant doctor in Trinity Bay, wrote to Governor LeMarchant in the fall of 1847 that people there still did not have enough potatoes to live on, and that supplies of potatoes were non-existent. Soaper feared that famine and disease would ravage Trinity Bay unless the government could find a way to deal with the potato crisis.[37] Again, government proved unwilling to spend money on relief. One correspondent, re-

flecting on the poverty felt in New Harbour at Trinity Bay, wondered
if government intended to ascribe to 'the Malthusian principle,' and
solve its problems by letting people starve to death without relief.[38]

The potato blight led one 'Investigator' to suggest that fishing families try planting grains instead of potatoes to provide for their own
subsistence. However, he cautioned against people thinking that agriculture could solve their problems: 'I am not one who dreams about
making this an agricultural country. With an immense and unrivalled
corn growing continent within a few days sail of us, it would be the
height of folly to attempt any separate division of labour of that sort
as to lead the people to expect that they would, or could, derive any
advantage from a competition with their more favoured neighbours.
As well might you attempt to establish a rival cod-fishery among the
Alleghaney mountains.'[39] Through 1848, as popular demands for relief
increased, the government sent out barrels of oats to see if this grain
might prove an adequate substitute for the potato. Unfortunately, unlike potatoes, oats would not thrive on the coast's soil and climate.[40]

While a generally poor agriculture worsened by potato blight drove
many to find their dinners in a heap of offal, government officials
were quite aware that credit practices in the fishery were the root
of the problem. When Poole merchants petitioned the British government for greater relief measures, Earl Grey, the British secretary of
state for the colonies, pointed out that it was the merchants who profited from the labour of fishing people, merchants who chose to make
themselves more secure by restricting the supply of provisions when
people needed them the most. Governor LeMarchant likewise found
that merchants were not much help in extending relief.

Unwilling to think that government might have to commit itself to
the long-term relief of fish producers to support the Newfoundland
trade, and unable to contemplate a fundamental reorganization of the
fishery to ensure that its returns went to producers rather than merchants, LeMarchant turned once more to the icon of agriculture. Diversification, the governor suggested, was the key to Newfoundland's
future prosperity. With some government support through potato seed,
road-work relief, and the work of the Agricultural Society, families
could be taught to look to their own resources to provide for their
own subsistence. Government relief and its encouragement of cultivation would serve as a means by which even more effort could be
squeezed from fishing families to keep the merchants in business.[41]
LeMarchant felt that his duty was clear. Government in Newfoundland

had to continue the policy of officially encouraging the cultivation which had begun with Governor Keats.

By 1848 LeMarchant had decided he would encourage the settlement of Newfoundland's interior. He suggested that Newfoundlanders experiment with better grains, fruits, and livestock breeds in order to find the best means of pursuing agriculture. To this effect he ordered a survey of lands in Conception Bay which might support the cultivation of wheat, barley, and oats. He also planned to give out some seed, but did not want to raise people's expectations that they would receive much government aid.[42] In short, LeMarchant thought that many of the plans put forward by the Agricultural Society were the key to Newfoundland's economic diversification.[43]

This renewed agricultural orientation by LeMarchant, like Harvey's before him, appears to have defied previous official experience with the northeast coast's limited soil and climate. However, the British imperial milieu may explain the governors' policies. A commitment to 'agrarian patriotism' dominated imperial officialdom of the period – a confidence that if colonies could only acquire the agrarian class structure of England through local agricultural improvement they could be integrated more easily into the structure of empire.[44] A spirit of improvement, marked by the commitment of gentry-backed agricultural societies to spreading English agricultural reform to allegedly ignorant and backward colonial farmers, was common in British North America during the first half of the nineteenth century. Under the patronage of colonial governments, which thought a better agriculture could revive ailing colonial economies, the societies had little actual impact on the manner in which rural households engaged in production in British North America.[45] Imperial sentiment, not a solid assessment of local resources, often lay behind agricultural schemes in the colonies.

While Governor Harvey had come to Newfoundland experienced in official beliefs about colonial improvement as a result of his tenures in Prince Edward Island and New Brunswick, Governor LeMarchant arrived fresh out of the military, with no colonial administrative experience.[46] LeMarchant himself made it clear that his commitment to Newfoundland's agricultural improvement stemmed from his belief that the impoverished residents of Newfoundland must look to their own resources, not those of the state, when the fishery could not provide their sustenance. The Newfoundland government had previously issued limited relief in the form of seed potatoes to northeast-coast communities, as a supplement to private charity and merchant credit, in the belief that it assisted only the most destitute, and would quickly re-establish fishing families' own cultivation capabilities.[47] But by the time LeMarchant became governor in 1847, the Newfoundland government could not keep up with the 'frightful' demands for relief by the island's residents.[48] The government's revenues could no longer support increasingly popular demands for relief, despite the years of encouragement to local agriculture.

Under LeMarchant's stewardship the Newfoundland government began to restrict relief expenditures on road-work paid in corn meal. LeMarchant declared that Newfoundland fishing families could no longer put their faith in government relief, but rather must look 'to their own exertions' alone in the 'increased cultivation of the Land' for sustenance.[49] After another summer of poor fishing and potato crops, the government declared that it simply could not provide any greater level of relief 'as it would be quite impossible for the most flourishing Revenue to sustain the majority of the population in which it is collected.'[50]

As an alternative to relief LeMarchant determined that he would spend money on model farms which would excite the people of the northeast coast to a spirit of agricultural improvement in the cultivation of grain and livestock.[51] When Robert Pinsent ignored the government's order to minimize relief to fishing families, LeMarchant reiterated to him that the Newfoundland government could no longer afford relief, and proclaimed that the colony must look forward to a day when it could supply all of the grain foodstuffs required by its population. Declaring that 'the idea formerly entertained of the utter barrenness of the soil is erroneous,' LeMarchant charted a course of agricultural improvement for Newfoundland similar to that which developed in the neighbouring British colonies.[52]

Outport people, struggling to survive the potato failures, could find little comfort in the Newfoundland Agricultural Society of St John's when it spoke of establishing model farms in the outports to show inhabitants how good agriculture might be practised. Observers in the outports felt that such schemes were a waste of time: 'for how is it possible that an indigent fisherman, without food, without clothing, without implements of husbandry, and under every other conceivable disadvantage, can follow the plans or imitate the example of those who have every convenience at their fingers' end, and who would have the best piece of land to be found in the whole locality at their disposal?'[53] Another correspondent scoffed at the society's plans to exhibit fattened purebred cattle. Reflecting on the cattle scheme a year later, this writer suggested that, for all the good it could do in an island where not enough crops could be raised locally to feed people, the Newfoundland government might as well send 'for 15 of the biggest Devonshire men to be found, high or low; every one of them with as many chins on him as there are spots on the nine of spades; and so distribute them about among the poorest parts of Newfoundland – as models for the inhabitants. Every man that *wont* [sic] grow in 6 months as fat as the Devonshire model, must get no more Government *male* [sic – meal]!'[54]

The *Weekly Herald* agreed with these criticisms, suggesting that fishing families would be better served by packing away fish offal during the season to help provide their winter diet. The paper advised people not to look to merchants for provisions on credit, when they threw away good food through the trunk-hole of a splitting table.[55] Again and again, from 1849 through the end of 1854, the potato crop and/ or the various fisheries failed. Time and again, the *Weekly Herald* provided the same suggestions as to how people should cope: use economy, eat fish offal, work hard, and make do on sparse government relief.[56]

While many fishing people did survive in these ways, Governor LeMarchant continued to support the Agricultural Society as the northeast coast hovered on the brink of famine. Ignoring the fact that outport inhabitants could hardly feed themselves when the potato failed, LeMarchant suggested that they should look forward to the day when a road might allow them to visit the Society's annual fall shows of 'stall fed oxen, fat sheep and hogs,' and perhaps compete for a prize at the yearly exhibitions. Instead of growing fat from looking at purebred cattle in their outports, people could come to St John's to do the same thing.[57] Rather than accept the responsibility for relieving those to whom merchants would not give credit, LeMarchant chose the fantasy of agricultural potential in Newfoundland: if the potato failed, bring in wheat, if that failed, then try barley or oats, and if they did not take to the climate or soil, then some type of better-bred livestock was the answer. There was always a disappointment for government in its agricultural policy, and always another cure-all. Government could not accept that Newfoundland's agricultural resources were at best only a poor supplement to the fishery because it would mean accepting ongoing responsibility for providing relief.

Many fishing people were not willing to live by eating refuse just so that they could supply salt cod for the merchants' trade. But in the face of the constant failure of agriculture all they could do was leave Newfoundland altogether.[58] Tired of the constant struggle to make a living in the fishery, many of the planters who still survived in the Labrador and seal fisheries began to consider the attractiveness of taking up farms in places like Wisconsin: 'numbers of persons – families as well as single men – are preparing to take their departure from this neighbourhood early in the spring; some of these are the owners of considerable plantations and tracts of land, and many of them we know to be in very considerable circumstances. Since the failure of the potato they consider it a hopeless task to contend with the arid soil of this country, while land requiring no manure and admirably situated for agricultural purposes is within so trifling a distance.'[59] Such fishermen had given up any hope of independence from merchants or of reducing reliance on mercantile credit for subsistence by resorting to cultivation, with its appalling record.[60]

It is not surprising that any fisherman who had experienced reasonable success and had a little property often found the prospect of emigration far more appealing than staying, and falling further in debt to merchants. Fishermen who had plans other than grubbing a subsistence from the soil could find no other alternative in Newfoundland. Any who could scrape together the money for passage fare left Newfoundland for the seemingly better prospects of owning a real farm in the United States. Some of these emigrants, such as Edward Pynn of Conception Bay, wrote letters to the *Weekly Herald* advertising their success in establishing near-200-acre farms, raising wheat and livestock. A Mr Hayward of Carbonear wrote to state that Newfoundlanders settled together in Washington County, Wisconsin, establishing their own family farms on which they could raise most of what they needed independently of any merchant and sell surpluses in exchange for goods they could not produce at home.[61]

The experience of Hayward and Pynn suggests that Newfoundland's resources did not allow fish producers to escape from dependence on merchant capital. The island did not possess the agricultural resource endowment which, in some other parts of America, proved to be the fertile soil in which industrial-capitalist social relations germinated:

America is an agricultural country, giving extensive employment to an endless variety of artisans in the manufacture of the raw material produced by different branches of cultivation, and so extensive as to afford an area amply sufficient for the investment of capital, and the development of industry and talent. There, no man need be idle who is inclined to labour, and all labour insures a reasonable remuneration. On the contrary, this Island can never become an agricultural settlement: here, no raw material is produced to call forth the genius, and reward the industry of the people, who are so pent up along the sea shore that the land already casts out its inhabitants. Besides, the employment generally is so connected with the sea that our native population know little or nothing of agriculture.[62]

Correspondents of the Harbour Grace *Weekly Herald* regretted that Newfoundland remained essentially a society based on the fishery because its resource could support little other industry. The colony remained 'in the hands of monopolists, who fix an arbitrary valuation on both exports and imports,' but no one could blame the colony's most successful producers for leaving. 'Alpha' wrote that merchants could make money off the trade in fish and oil, provisions, and goods, but that merchants were, like most capitalists, in the business for their own profit, not the welfare of the community. For the actual catchers of fish Newfoundland provided little means of improving themselves, other than its fishery. The airing of such views in the press indicates that people were aware that life was not likely to get much better in the fishery as long as fish merchants dominated it. But only government had the power to initiate meaningful change, and it was not about to tamper with the class organization of the industry. Government's only other alternative was to promote farming.[63]

When planters began to leave Newfoundland in the mid-nineteenth century, their actions served as a mute testimony to the futility of government's embarking on agricultural development policies as a means of answering the constant provisions crises arising from the fishery. In the first years of representative government, Governor Cochrane steadfastly resisted reformer rhetoric about the bounty of Newfoundland's soil and climate. Yet ongoing depression in the fish trade encouraged fish merchants to restrict credit to fishing families for essential food. Unable to find any meaningful substitute for their subsistence in local agriculture, especially with the potato failures of the 1840s, fishing families turned to the state for relief to stave off famine. Unwilling to accept the burden of long-term relief expenditure, and unwilling to contemplate restructuring the fishery in any way that would free families from reliance on merchants' imports of food, successive Newfoundland governors turned to agriculture in the hope that it would provide an alternative to government relief. That hope could not be sustained by Newfoundland's resource endowment, but rather only by government embracing the reform chimera: agricultural backwardness persisted because merchants opposed farming.

8

Liberals and the Law

Reform prevarication about Newfoundland history extended to much more than agriculture. The Colonial Office granted representative government to Newfoundland, but not responsible government which would allow the growing bourgeoisie of St John's to control colonial revenue and patronage. As governors became as enthusiastic about farming as reformers, the latter group needed another issue in order to build broad popular support for their political demands. Recognizing that fishing people were a potent force in the shaping of Newfoundland's politics and society, but a force tied to the unequal accommodations implicit in truck, reformers changed tack in their assault on government by inventing a new history of the wage and lien system. They now suggested that this system was a custom of the resident fishery which benefitted planters and supported their employment of servants. The reformers, now known as Liberals in the House of Assembly, did this to attack the Executive Council, particularly their early nemesis, arch-conservative Henry Boulton, chief justice and president of the Executive Council, who decided that the wage and lien system should not be revived after the temporary law of 1824 (which extended the life of the wage regulations in Palliser's Act) lapsed in 1832.

By identifying Boulton's actions on behalf of fish merchants' interests, Liberals struck at the paternalist bonds which tied fishermen and merchants together in northeast-coast society. Boulton's actions were a perfect occasion for Liberals to build popular agitation for more political reform around the 'outrage' of an outsider arbitrarily using his authority to overturn a long-standing 'custom' of the fishery, although previous chief justices had ruled that it was more accurately a custom of the old migratory fishery only. In doing this, Liberals mythologized the wage and lien system as something necessary for the prosperity of planters and fishermen alike.

The war cry of 'outrage' served as a Liberal means of drawing the producing classes into what was otherwise a largely sectarian-influenced but elitist political struggle over constitution and patronage, which remained somewhat esoteric to the day-to-day struggle for survival of outport fishing people. As in the other British North American colonies, real political power in Newfoundland lay with the executive. Governors controlled appointments to their councils, usually choosing local military and government functionaries rather than members of the colonial bourgeoisie, whether the latter were the great fish merchants or the smaller trader professions. Governor Cochrane and his successor, Prescott, particularly distrusted Liberals such as Patrick Morris and William Carson, not only for their criticisms of merchant influence and official Protestantism in imperial policy, but also because of their willingness to exaggerate the developmental potential of Newfoundland. Liberals, with the notable exceptions of Carson (and later Robert Parsons), were predominantly St John's Roman Catholics who enlisted the aid of their bishop, Michael Fleming, in their bid to use responsible government to break governors' tendencies to patronize Anglicans and merchants with government appointments and favours.[1]

The governor, Sir John Harvey, subsequently managed to co-opt old reformers like Morris and John Kent during the 1840s, by patronizing agriculture and the Roman Catholic hierarchy in order to calm a rising tide of sectarian violence in St John's. His successor, Governor LeMarchant, encouraged by a Tractarian Anglican Church hierarchy bent on depriving Catholic and Wesleyan Methodist schools of public funding, distrusted the Roman Catholic influence in the St John's bourgeoisie, and reverted to older patronage styles. The Liberal party rekindled under its leader, Philip Francis Little, and its propagandist, Robert Parsons, as the plums of patronage slipped from their fingers. Through the late 1840s and 1850s, Little's Liberals refined their demand for constitutional change by demanding a responsible party government. Governors, they said, should have to choose their executives from whichever party elected the most members to the House of Assembly.[2]

Parsons played a crucial role in the Liberal reinterpretation of the wage and lien issue. A Presbyterian, he could not wholeheartedly accept Irish Catholic sectarianism as the basis for Newfoundland Liberal ideology. He preferred a more classical Liberal assault on the arbitrariness of authority, and as editor of the St John's *Patriot*, focused on Chief Justice Boulton. At first unfocused, this attack narrowed in on the wage and lien issue, supplying Liberals with a cause to replace the platform lost to Harvey's and LeMarchant's patronage of agriculture, and to provide a basis from which to question the fairness of government dominated by the 'mercantocracy.'[3]

Much planter and merchant sentiment, ironically, grew against the wage and lien system during early investigations of constitutional provisions which might replace the temporary Fisheries and Judicature Acts of 1824. To get recommendations about what new laws for the colony should look like, Cochrane, in 1829, had sent notices to various parts of Newfoundland that the magistrates were to organize community meetings to deliberate upon the matter. Suggestions poured in from all over the northeast coast. In Conception Bay, where local merchants supported reform demands for self-government, feelings ran high against both acts. Public meetings of planters and merchants produced declarations of support for a local legislature to make laws suitable to a mature resident fishery. Participants tended to feel that the circuit courts were better than the Surrogate Courts, but wished for laws which would impose greater discipline on servants and abolish the law of current supply. Outside Conception Bay, in areas still dominated by English merchant houses, planters and merchants disagreed. At Bonavista, Greenspond, and Twillingate, planters and merchants wanted to give current suppliers precedence over servants in prior claims of the produce of fishing voyages. Planters needed not only freedom from servants' wage liens, but also greater legal penalties for servants' negligence. Such demands reflected the need to give merchants and planters in more isolated locales extra security for credit employed in the fishery.[4]

As the British government drew closer to having to make new constitutional provisions for Newfoundland, in a report to the Colonial Office the judges of the Supreme Court in 1831 (Chief Justice R.A. Tucker, A.W. DesBarres, and E.B. Brenton) identified the law governing masters and servants and current supply as their most pressing problem. The judges noted that current supply was a usage which had been derived by merchants who made the transition from a migratory to resident fishery; that it only later became sanctioned by law. Merchants never accepted the preferential claims of servants for wages, believing that it encouraged servants to work hard only until their own wages were covered by the planters' voyages. As a consequence, merchants stretched the meaning of current supply far beyond its original meaning. It had come to apply to anything a planter or fisherman took on credit in a year, not merely to supplies specifically required during and for the fishing season. This put a lien on all planters' production in the current year, allowing planters little leeway in capital accumulation. The Supreme Court judges saw this as a departure from the original custom of the fishery; they noted that, since the time of Chief Justice Forbes, the Supreme Court had always tried to reassert the original usage of current supply.[5]

Tucker, DesBarres, and Brenton felt that the time had come to set labour and capital in the market of the Newfoundland fisheries free from the restraints of the wage and lien system. Dismissing English merchants' claims that the end of current supply would see merchants withdraw their capital from the fishery, the fishery ruined, and fishermen starved, the justices recommended abolishing the laws of wage preference and current supply, feeling that the fishery could only benefit from this change. If their suggestions became law, the judges felt that merchants would only advance credit to planters who paid the debts of previous years. Without the law of current supply, planters would have to operate a profitable fishery from year-to-year if they were to secure merchant credit. Servants would have to work harder to ensure the planter's profit, as a lien would no longer give privileged security for their wages.[6] Deprived of easy credit under current supply, planters who could not succeed even with a change in wage law would have no choice but to work for their neighbours. Consequently, successful planters would be able to profit from their fellows' failures 'just as Pharaoh's lean kine ate up the fat ones.'[7]

In 1832 Attorney-General James Simms recommended similar changes to the laws governing the fishery of Newfoundland. Simms felt that the fishery was too complex to be governed by any one custom or set of laws. He believed, as former Chief Justice Forbes had before him, that the best law for Newfoundland was the flexibility of English common law; that planters should be able to discipline their servants more severely to ensure greater productivity. Simm's planter bias emerged in his recommendation that desertion be made a criminal offence punishable by a prison sentence. Like the Supreme Court judges in their 1831 report, Simms felt that the wage and lien system was not a set of customs either stemming from or suitable to the resident fishery. Planters, he said, should have to stand on their own without the prop of current supply, or the ability to foist responsibility for servants' wages onto the backs of their merchants. As a result, class differentiation based on a more sound capital accumulation would begin among the planters, as those who could not survive except by the artificial means of the wage and lien system disappeared.[8]

The British government relieved itself of the burden of dealing with the wage and lien issue in 1832 by granting Newfoundland a representative government with full powers to legislate in matters of wage and credit law. Within the House of Assembly, Liberals forced Chief Justice Tucker to resign in their fight over money bills. In 1833 Lord Stanley, the British colonial secretary, appointed in Tucker's stead Henry John Boulton, who had been dismissed as Upper Canada's attorney-general shortly before for his role in the Upper Canadian Tories' fight against the reformer, William Lyon Mackenzie.[9]

Boulton, as the attorney-general for Upper Canada, opposed reformers who gained much popular support from those who felt that a Tory-dominated bench did not administer an equitable justice. Opponents of the Upper Canadian executive seized on a number of court cases, usually termed 'outrages,' in which Crown officers could be seen as denying the due process of law. To most Upper Canadians it was not so much that authority perverted the law to repress people politically, as that a supposed clique of merchants and government officials, 'those whose aim in life was to make a fortune,' used the law to exploit a society of agricultural petty producers, 'those whose main aim was to make a living.' Boulton became embroiled in a number of scandals which, to many Upper Canadians, served as examples of how the rich used the courts for their own benefit. Upper Canadian reformers increasingly concentrated their attacks on Boulton's misuse of his office. Boulton responded by fighting a pitched battle with reformers in the legislature until his expulsion.[10]

In his new appointment Boulton felt compelled to bring both Newfoundland criminal and civil law in line with what he understood to be common practice within the British Empire.[11] As president of the Executive Council, he resisted Newfoundland Liberal demands for government reform just as he had that of the Upper Canadian reformers. When Newfoundland Liberals began to attack Boulton as part of their assault against executive power, he used his position as chief justice to persecute them. His opponents accused him, probably justifiably, of arbitrariness.[12] Under the leadership of Morris and Carson, Liberals were not long in seizing the advantage presented to them by the chief justice. In 1835 they engineered the writing of a petition which accused

Boulton of anti-Roman Catholic 'bigotry, illiberality and intolerance.'[13] Liberals accused Boulton of partiality in supposedly favouring merchants over fishing servants by throwing out of council a reform-sponsored bill which would have resuscitated the wage and lien system, dead since the expiry of the 1824 acts.[14]

Liberals used Boulton's interest in legal reform to appeal to the northeast-coast's producing classes, particularly fishing servants. The message was simple: Chief Justice Boulton was an Upper Canadian Tory bent on denying fishermen their ancient custom of having a lien on the supplying merchants for their wages, thus causing their families' starvation. In yet another petition, in 1837, they claimed that the chief justice transgressed against 'the rights and privileges of the people.' He had violated the rule of law as a stranger to Newfoundland, ignorant of its fishing customs, and dismissive of opinion more familiar with the colony's laws.[15]

Liberals in Newfoundland, like reformers in Upper Canada, used legal 'outrage' as a rallying cry in their struggle against executive authority. Yet in this case Boulton committed no legerdemain in the wage law issue that personally benefited him. Instead, the chief justice stood squarely behind the principle that the law should no longer inhibit free exchange between labour and capital in the market. The wage and lien system did not benefit planters, but reformers could create an image of a partial justice undercutting producers' 'rights' through manipulation of the laws in favour of merchants. Reality did not matter as much as illusion in this political struggle.

In Upper Canada the Tory factions to which Boulton belonged could use the issue of loyalty to cultivate support among members of the producing classes. But when Boulton came to Newfoundland he found no society of Irish Protestant, British Protestant, and Catholic farmers and mechanics, all willing to show their support for the Crown. Instead he found many who were willing to show their dissent. Neither did Boulton find the equivalent of Upper Canada's Scottish Catholic bishop willing to stand with the Executive Council in the hope of gaining official patronage for his church.[16] The early Liberal movement of which Boulton ran afoul was predominantly Irish Roman Catholic – except for Carson and Parsons – with its own ethnically and religiously coloured grievances, and Boulton did not understand the new political ground on which he had to fight.[17]

Most of the immigrants trickling into Newfoundland after 1815 were Irish Catholic servants. While the first Roman Catholic bishops tended to support the governor and his advisors, Bishop Fleming swung behind the Liberals from 1830 in an effort to gain more patronage for the Catholics and state support for separate Roman Catholic schools. As part of his fight against the Newfoundland government, Fleming joined the Liberals in condemning Boulton's attempts finally to end the wage and lien system, a supposed fight for servants' 'rights' against the merchants. Fleming built on a strong Newfoundland tradition of itinerant plebeian priests such as Patrick Power, who often acted without the sanction of hierarchical authority. These priests led their largely servant flock, of the same Irish background, in faction fights to keep wage rates up by fighting off competitors from other Irish groups for jobs in the fishery.[18] Fleming also brought with him from Ireland the tactics of the clerical supporters of Daniel O'Conell. In the O'Conellite tradition priests would fight for the rights of the indigenous Catholic bourgeoisie against the Protestant Ascendancy by deflecting the discontent of their fellow Roman Catholic labourers and tenants at social and economic inequality. The discontent of the Irish producing classes consequently fuelled nationalism rather than class consciousness, and nationalism and religion ran parallel. In Newfoundland the rising St John's shopkeeper bourgeoisie and their outport allies (many of whom, like Morris and John Kent, were Catholic) pursued a similar means of cracking what they saw as an oligarchy of English bureaucrats and merchants governing the colony.[19]

The Irish servants who supported the Liberals did not simply do the latter's bidding. Servants allied with them to strike at what they saw as exploitation of their labour by merchants' truck. The power of fishing servants' riots on behalf of Liberals during the elections of the 1830s was a stern reminder to the latter that servants must be courted, that they would not give blind support. When servants backed the Liberals they did not fight simply for Catholic rights, but for protection of their own interests. Liberals took great pains to identify such interests with the issue of the wage lien rather than the real problem of merchant domination, in an industry which proved able to support only merchants or residents satisfactorily, but not both, under its present organization.[20]

Boulton defended himself against his critics at length, by responding to Bishop Fleming's 1835 defence of 'fishermen's rights' against the chief justice's court rulings.[21] Boulton outlined three court decisions in which he decided on wage law. The first involved a Ferryland servant, Thomas Reilley, who sued his master, planter Richard Sullivan, and his master's supplying merchants, Codner and Jennings. The second was between a fishing servant, Silvey, his master, Morgan, and their supplying merchant, Bennett. The third was the most controversial: *Colbert v. Howley*. In all these cases Boulton would not support servants' wage liens on fish in the hands of supplying merchants because he could find no formal contract between merchants and servants for employment in the market. Servants could hold their masters, the planters, liable for wages, but not merchants.[22]

Boulton charged the jury in *Colbert v Howley* that Colbert had contracted as a servant with planters Grant and Hamilton who returned only £44 in fish and oil against £160 in credit advanced by their supplying merchant, Howley. There was no evidence to suggest that Howley took any responsibility for Colbert's wages, although the merchant had received some fish and oil from the planters. Yet Colbert sought full payment of his wages from the merchant. The chief justice argued that no custom of wage lien could be proved to have been generated out of the resident fishery. Such a lien had been enshrined in the now defunct acts of 1824, and had been included there as a custom extending out of Palliser's Act and the migratory fishery. Boulton could find no consistent statement of present usage in the fishery, but all witnesses were consonant in that 'no one pretended that the merchant was liable in the first instance, and without reference to the master or planter, as he must be to be subject to an action at Law.' Boulton argued that the 1824 Fisheries Act justly gave fishermen and seamen a lien for the payment of their wages or shares against the employer, the planter.

Such a lien was quite in line with English law, but to extend it to the people the employer had dealings with, went too far. The chief justice, like Tucker, Brenton, and DesBarres before him, felt that the court's duty was to resist any further pressure to shape law to fit this supposed custom. In his capacity as president of the council, Boulton stated that the executive furthermore would not consent to any new fishery bill because the House of Assembly could not agree as to what the customs of the resident fishery were; and that he therefore could not base judgments on unconfirmed customs.[23]

Boulton thought that the justification for his decision lay in the very testimony offered by both the plaintiff's and defendant's witnesses in the case of *Colbert v Howley*, including prominent Liberals such as Morris and William Thomas, but his opponents turned the case into something of a show trial as part of their effort to discredit the chief justice.

Thomas, a St John's merchant, stated that Palliser's Act created the wage and lien system whereby 'the Merchant received the fish & oil subject to the payment of the Servants wages out of the proceeds of the fish & oil.' Despite his interest in defending the wage and lien system, Thomas could outline no consistent use of it. Even he did not always honour a servant's lien on the fish and oil Thomas received from outport planters, but appeared to do so only when he had personally approved written wage agreements planters had made with servants. Furthermore, merchants only accepted servants' liens when they had received all of their masters' produce. Otherwise a servant had to track down all the merchants his master dealt with, and sue each for wages in proportion to the amount of fish and oil they received. All that Boulton could garner from Thomas's testimony, which Morris repeated, was that sometimes supplying merchants would pay wages to the extent of the fish and oil received; and the assertion, without proof, that the wage and lien system was a long-standing, convoluted custom of the resident fishery. Furthermore, supplying merchants usually were privy to the number of servants the planter hired before they issued supplies. Thus, despite his attempt to reinforce the Liberals' support of Colbert, Thomas's testimony only told Boulton that merchants were never held liable for the full payment of servants' wages, and that merchants only paid wages when they were privy to a contract between planter and servant. Boulton heard the testimony of others, including supplier John Brown, supplier James Fergus, servant John Cuddahee, and Francis Tree, all of whom asserted that the wage lien had always existed as a custom of the fishery.[24]

The chief justice's refusal to accept that servants had a lien against supplying merchants, based on such contradictory and suspicious evidence, provided the basis for a Liberal declaration that such arbitrary justice was an outrage against fishing servants. After Boulton announced his decision, William Carson, Patrick Morris, John Kent, and J.V. Nugent led an open-air protest at St John's. This demonstration resulted in a petition against the chief justice outlining the Liberal grievances. The petitioners accused Boulton of perverting the Newfoundland justice system by ignoring its long-held customs. The justice system had been altered by a person 'who came to this country with the character of being rancorously opposed to the liberties of the people.' They accused him of being immersed in anti-Catholic party politics and favouring fish merchants over servants by striking down the wage lien. Finally, as president of the council, Boulton had led their fight against Patrick Morris's attempt to introduce a new bill to effect such a lien. This, charged the Liberals, demonstrated that Boulton bore a 'rancorous hostility to the interests of the poor.'[25]

The image of Boulton as the oppressor of the poor emerged even more strongly in the editorials of Parsons' The Patriot. The Royal Gazette defended Boulton as a protector of property. The Patriot responded with an aggressive, melodramatic assassination of Boulton's character and his jurisprudence. Parsons claimed that Boulton's legal decisions enslaved servants and took food from the mouths of babes: 'Thus has Boulton's law made us a pauper population – a pennyless people – and put the just dues of the Fisherman and the Shoreman into the pockets of the Merchant!' The Liberals worked to create two popular images. The first was of Boulton as a foreign, despotic magistrate who capriciously overturned previous Newfoundland justices' rulings in the courts. In the second, Liberals alluded to the chief justice as the fish merchants' man, ruling over Newfoundland's poor without care or feeling. Throughout, the paper's message linked Boulton to a general problem: Newfoundland's not being able to govern itself, and being subject to a justice system imposed on the colony from London.[26]

The Patriot cited two previous chief justices, Forbes and Tucker, as being near-heroic defendants of the fishermen's rights. Forbes was the judge who in 1816 waved 'the magic wand of the Enchanter' to make Newfoundland's justice system subservient to the interests of fishermen. The paper suggested that 'the benevolent Tucker, the Poor Man's Judge,' confirmed fishermen's constitutional rights in the wage and lien system, and Forbes's general regulations for the legal system. This legal idyll continued 'until the ex-Attorney General of the Canadas' overturned their decisions: 'the Charters of the country were set at nought, and the laws outraged.'[27]

Newfoundlanders, declared the Patriot, must stand behind the Liberals' petitions against Boulton, and for 'the restoration of Tribunals that, by the Constitution of England, are intended to be the Protectors of the lives, the liberties, and the properties of her subjects.'[28] Correspondents agreed with the paper, contributing to the myth that Boulton's rulings were an abrupt break with the former chief justices' decisions. One 'L' suggested that Boulton's ruling, in the case of Colbert v Howley, had taught fishermen that, until Newfoundland was rid of Boulton, 'the merchant may riot in the spoils of the poor – the servant must starve in silence and content.'[29]

Parsons reproduced his version of the minutes of *Colbert v Howley*, punctuating them with compliments for Thomas and Morris, and portraying Boulton as the epitome of arrogance. The essence of these minutes did not materially differ from the chief justice's own, except for the paper's editorializing. Again, the *Patriot* took great pains to identify Boulton with all opposition to constitutional reform, particularly responsible government, and with the commitment of an arbitrary act against justice. In reporting that Boulton instructed the jury to find for Howley by saying, '"AND IF MY HEAD WERE TO BE CUT OFF, I COULD NEVER BE MADE TO GIVE ANY OTHER DECISION,"' the editor claimed that the chief justice had unfairly predetermined the jury's verdict in a perversion of justice: 'and thus was a premium, a bounty given for the commission of crime, for the perpetration of outrage, by those whose duty it was to throw the shield of the laws round the oppressed – to protect the poor from the rapacity of the rich.'[30]

A correspondent of the conservative *Public Ledger* could not let the Liberal attack on Boulton go unchallenged, noting that the Liberals were on shaky ground claiming that the wage lien against supplying merchants was an 'ancient custom' of the fishery. The lien, to be considered a custom, would have to have been in use so long that no contrary memory of other practices existed, that the use have been continuous, that it was reasonably and peaceably accepted by society, that the custom be certain, that its observance be compulsory, and that it was consistent with other customs. The correspondent, 'One of the Natives,' cited a letter by George Larkin to the British government in 1703, which observed that merchants habitually carried off planters' fish without paying servants' wages. No lien existed in custom at that time. The British government legislated the wage lien into existence to confirm governors' declaratory efforts to stem planters' connivance with merchants to pay their accounts before wages.[31]

'One of the Natives' pointed out that court decisions based on the wage lien law were not to impose a complete lien for all servants' wages supplying merchants as such, but only to attempt to follow the effects of insolvent planters as far as they could in securing the wages of such planters' servants. Servants had liens only on their masters' fish, not more general ones against supplying merchants. Merchants had to pay wages out of the proceeds of the sale of the fish and oil the planters gave them, and no more. Palliser's Act confirmed that servants did not have to retain the actual fish and oil to enforce their lien on merchants to this extent, so that merchants would be able to receive produce freely for marketing throughout the season. Again, legislation not custom underwrote the wage lien, and British officials had no intention of allowing the wage lien to govern the resident fishery at Newfoundland. The demise of Palliser's Act and the 1824 Fisheries Act suggested 'One of the Natives,' finally ended the legislative basis for any wage lien against merchants.[32]

In 1836 Liberals again sent a petition to London complaining about Boulton, and demanding that the Colonial Office investigate him. This further entrenched in the public mind that they led the fight for servants' wage 'rights.' To rally fishing servants to their cause the Liberals held public meetings at Harbour Grace, Brigus, and Carbonear. Citizens' Committees called together the servants and circulated petitions among them to sign, condemning Boulton.[33]

Through 1836 the Liberals continued to suggest that Boulton stood for tyranny and the denial of every British subject's constitutional right to a responsible government. The *Patriot* was open in identifying the Newfoundlanders with Upper Canadian reformers and demanded some form of united front.[34] Correspondents demanded that Boulton, the oppressor of 'the poor Irishmen' and 'the lower classes,' be opposed by all planters and fishermen.[35] When merchants petitioned on behalf of Boulton, The *Patriot* argued that their petition could only be secured as 'servants have been threatened, and labourers seduced, and young boys *bought*, and the Sealers offered to be bribed, and the foreign crews of merchant vessels humbugged to append their names'[36]

Morris and Parsons constantly referred to Boulton as a tyrant, a likely follower of Upper Canada's reactionary governor, Sir Francis Bond Head, 'the redoubted Tory of Toronto.'[37] Liberals, on the other hand, were 'the friends of the King and Constitution, of Chartered Rights, of Liberty, of Civilization, of Equal Laws and Justice. In fact, The People.'[38] When Sir Francis Bond Head used extra-constitutional means (such as the intimidation of voters by his Orange Order supporters) to win the Upper Canadian general election of 1836, and a subsequent legislative investigation whitewashed the affair, The *Patriot* claimed that the government's refusal to get rid of Boulton amounted to the same thing. 'Has not this Colony of Newfoundland experienced a similar outrage, again and again repeated as that which has just been perpetrated upon the Province of Upper Canada?'[39]

Asked by the local colonial secretary to state exactly how he would see constitutional reform proceed on the issue of wage law, Boulton

at this → Boulton little a merchants a

waved a red flag in reformers' eyes. He argued that Liberals, rather than planters, became the advocates of a servants' lien against merchants because they simply wanted to stir up the producing classes to support their quest for responsible government. Noting recent election riots in favour of reform candidates in Conception Bay, Boulton suggested that Newfoundland society was too susceptible to the Liberals' tactics. Without an agriculture capable of supporting a proper gentry, Boulton felt that the dichotomy of Newfoundland society, between merchants and fishermen, created a volatile political climate that could be ignited quickly by Liberals, 'who chiefly attain notoriety by keeping the lower orders in a state of constant excitement.'[40]

Boulton furthermore did not blunt his verbal reprisal against reformers' attacks on him and his legal decisions. During the summer of 1837 he wrote to the governor that the Liberals had allied themselves with an insubordinate Roman Catholic Church, and were willing to use violent intimidation to attain their goals. Boulton proudly cited their petitions against him in his own defence, because they showed that the chief justice had 'an unceasing and uncompromising opposition to the encroachments of unauthorized power upon the legitimate rights of others.' The reformers, claimed Boulton, gave him an undue importance as the sole opponent of the wage and lien system, unfairly singling him out when the entire conservative legislative council opposed it. Boulton was right in identifying the reformers' tactic as making him a symbol of tyranny over the wage issue in the minds of fishing people. But the chief justice's language made him an easy target, especially as Boulton made no secret of his dislike of the Newfoundland Roman Catholic establishment.[41]

Boulton's further legal reforms provided the Liberals with a new series of 'outrages' to use in their complaints to the Colonial Office. In 1837 the reform-dominated House of Assembly struck a committee to investigate Boulton, and sent its report to London. Besides altering the jury system, Boulton, in 1833, had changed the writ of attachment issued in civil cases. The altered writs allowed creditors greater ease in securing their debts from defaulting accounts by removing an exemption which protected from attachment all property essential to the fishing season. This change in the writ was in keeping with Boulton's belief that the law of current supply harmed the fishery. The report noted that people of capital no longer involved themselves directly in the production of salt cod, but rather did so by advancing those without 'money or property,' boats, nets, lines, provisions, and clothing to make a voyage. The old writ of attachment meant that a current supplier did not hesitate to issue such supplies because he did not have to worry about creditors' suits against planters or fishermen from the previous season. Liberals condemned Boulton for this striking at current supply, a 'custom of the country' from 'time immemorial.'

As well as current supply, the reformers included another protest against Boulton's attempt to declare that a preferential lien no longer existed for fishing servants' wages. The reformers claimed they had 'the best and most authentic documentary evidence before them to prove' that the wage and lien system was a custom of the fishery, but could only produce late-eighteenth-century governors' proclamations securing fishermen their wages. While they were complaining about Boulton, Liberals sent other messages to London calling for more local legislative control over the executive to avoid such arbitrary use of power in the colony's government.[42]

In 1838 Patrick Morris charged that the chief justice had done nothing less than sweep away *en masse* the whole of the laws, usages, and customs, which for centuries regulated the trade, fisheries and industries of the Island of Newfoundland.'[43] Boulton, according to Morris, had ignored the precedents set by former chief justices in sanctioning the wage and lien system, and was ignoring the maritime law of Britain in which it was rooted: the law of Bottomry, which gave the last supplier of necessaries for a ship, a preferable claim over all former suppliers. Maritime law protected not only current suppliers, but also the preferential claim of fishermen to their wages.[44]

Boulton had not disputed that fishermen had a lien against their masters, the planters. In the case of seamen, as Attorney-General James Simms later explained, they too had a lien against their employer, the owners of a vessel and their representative, the ship's master as manager of the owner's capital and hired labour. In 1837 the governor gave his assent to a bill, giving Newfoundland seamen a lien for their wages against shipowners (1 Victoria c. 9).[45] But in the case of fishermen, merchants did not hire their masters – the planters – nor own the planters' capital; thus, fishermen could not proceed with a lien similar to that of seamen.

Morris was willing to falsify the past to attack Boulton. Eighteen years earlier the naval Surrogate Courts had been reviled by Carson and Morris when Surrogate judges had ordered the whipping of Butler and Lundrigan. The reformers used the Surrogate Courts then as 'an image of official tyranny over poor, helpless, outharbour fishermen ... to galvanize the public into recognition of the need for judicial and constitutional change.'[46] Now it was convenient for Morris to portray them as part of an old lineage of judges who had protected the fishermen's rights which Boulton now sought to undermine:

It was the invariable policy of England to watch with the greatest attention over the fisheries of Newfoundland, it was called a nursery for seamen, and the parent government watched with more than the care of a nurse, the interests of this invaluable class. By a reference to the history of the government of Newfoundland, it would appear that the sole object of Government, Governors, Surrogates, Courts, and Judges, and all, was to protect the fishermen and seamen from the oppression and injustice of the merchants. An uninterrupted, interminable war has raged between the government on the one hand, and the merchants on the other on this very point.[47]

In yet another misrepresentation, Morris claimed that the custom of current supply stemmed from the custom of merchants extending winter supplies on credit against the next year's voyage. This explained how the custom was extended to cover an entire year rather than just for the fishing season. Morris did not acknowledge that merchants had been unwilling to grant winter supply on a customary basis since at least the decisive shift to a resident fishery after the provisions crisis of 1816–17.[48]

Besides revising the history of surrogates and winter credit, Morris did not accurately reflect the views of the past chief justices when he declared that they supported the wage and lien system. Besides the clear dissent of the 1831 Supreme Court report, Chief Justice Tucker and his predecessor Forbes both felt that the system was ill-suited to the Newfoundland fishery. Forbes had felt that the courts wrongly extended wage and current supply liens to cover the year-round needs of the resident fishery, and he gave a detailed description of how injurious the system was to capital accumulation while stating that he would accept the perversion of current supply because it would not be politic to do otherwise. Palliser's Act had become bound up in the general rule of law and could only be corrected by legislative, not judicial, decision. As all creditors waited until planters could satisfy wage claims, the definition of current supply was effectively extended beyond its original meaning. Chief Justice Tucker had agreed with Forbes' ruling about the inappropriateness of the wage and lien system as it operated in the Newfoundland fishery. In an 1823 case Tucker declared that merchants used the lien of current supply to support planters whose unprofitable activity would otherwise end their enterprise.[49] Tucker felt that Newfoundland law had not sufficiently protected the resident fishery from the incursions of the wage and lien system since, and his dissatisfaction found itself in his 1831 recommendation that the system be abolished entirely.

Boulton travelled to London in 1838 to defend his legal decisions. He maintained that he bore no ill will towards fishermen, but reiterated that he could not support the Newfoundland wage and lien system, citing in his defence both the report of Justices Tucker, Brenton, and DesBarres in 1831 and the Simms report of 1832. Boulton denied ever addressing the current supply issue directly in court – he had done so only indirectly by altering the writs of attachment – but claimed that previous creditors could not have the security of their investments superseded by current suppliers' extending even further credit to planters of dubious backgrounds, yet propped up by the law of current supply.[50]

The chief justice's defence convinced the Colonial Office that his legal decisions were sound, and it saw no reason to overturn them. But Lord Glenelg, the undersecretary of state for colonial affairs, decided that Boulton had made himself too unpopular with Newfoundlanders. The chief justice, unlike the reformers, did not understand that political leaders maintained much of their authority through the cultivation of producing classes' support. Boulton's battles with Bishop Fleming and prominent Catholic Liberals, combined with his patrician disdain and arrogant tone in dealing with any issue, made him unpopular generally, but particularly with the great mass of Catholic fishermen in Conception Bay. In the summer of 1838 the Colonial Office appointed a new chief justice, John G.H. Bourne. Boulton returned to private practice in Upper Canada.[51]

Throughout the 1840s Liberals either introduced or supported various bills which attempted to declare the existence of some form of wage and lien system. All of these bills fell prey either to internal Assembly disagreements or to opposition from the Executive Council.[52] Boulton's successor, Chief Justice Bourne, saw it as his duty to continue resisting any attempt by reformers to see a lien realized against supplying merchants for servants' wages. Bourne ruled, in *Nowlan v McGrath*, servants could not stop planters from delivering fish and oil to their merchants if the servants anticipated that the planters would have trouble paying wages. Such actions by servants constituted a form of insubordination during the tenure of their service, and could not be tolerated by the courts. Following what The *Patriot* claimed was simply the 'dark' precedent of Boulton's rulings, Bourne ruled that if servants agreed to season-long contracts, they could not interfere with their masters' business until the contract expired; that the sanctity of the marketplace must be observed for the orderly conduct of business. Servants' only redress was to negotiate the payment of monthly wages – an impossibility in an industry in which planters only realized their gain (if any) at the end of the fish marketing season.[53]

In 1841, in an attempt to lay to rest the controversy over the wage and lien system, the judges of the Supreme Court issued a report on the issue. Two, Bourne and Lilly, accepted that no lien existed, while Assistant Judge DesBarres, who had run afoul of Boulton in the operation of the courts, abandoned his previous agreement with this position, and sided with the Liberals. DesBarres tried to avoid giving any opinion on the law by claiming that the statute law on the matter had expired and could only be restored through new legislation, and therefore, that legislators should be making new laws rather than have judges commenting on old ones. Having said this, DesBarres claimed that the wage lien did exist. The editor of the *Public Ledger* wondered what DesBarres could be basing this opinion on, as he conceded that there was never such a lien based on custom, and its existence in statute had expired.[54]

The Newfoundland governor's law officers did not accept that the law of wage and lien continued to exist. Attorney-General Simms stated unequivocally that Palliser's Act had entrenched a migratory fishing law as a perpetual act, and that British officials had extended its rule over the resident fishery by the temporary Fisheries and Judicature Acts of 1824, until such time as the Newfoundland government could make a law more suitable to the resident fishery. The 1824 Fisheries Act had repealed Palliser's Act, so that the former's expiry in 1832 had made the preferential lien for fishermen's wages a dead letter. Solicitor General H.A. Emerson agreed with Simms.[55] Throughout the late 1830s and 1840s, then, servants had a lien against fish and oil only in the hands of their direct employers, not in those of the fish merchants. Servants could take no action to secure wages if they interfered with their masters' marketing of fish and oil. Merchants no longer had a lien on planters' fish or effects as current suppliers. Old debts took precedence over current ones in the settlement of insolvent estates.

As a result of Simms's and Emerson's 1841 decision, the Liberal-dominated assembly again tried to introduce a fisheries bill which would restore the wage lien against supplying merchants. The bill demanded that the government declare this lien a custom of the fishery, one which allowed fishing servants to follow the fish and oil they caught into the hands of merchants. The Liberals departed from the old wage law of Palliser's Act by proposing that penalties for servants' negligence or absence be vastly increased, from two days' to twenty days' wage forfeiture for every day of neglect. The government refused to allow the bill to pass into law because it first demanded the declaration of a custom which did not exist, and then actually submitted fishing servants to greater oppression by their masters, with greatly expanded penalties for negligence. The provisions did not specify exemptions if absences were caused by something like illness.[56]

Robert Parsons, who briefly flirted with a merchant-Protestant liberal alliance from 1840 to 1842, demonstrated his own lack of a sincere commitment to the wage lien by opposing any attempt to legislate a new one.[57] Boulton and his successor had done the dirty work in securing market freedom in labour and capital, and now the *Patriot* was willing to let the issue rest on fishermen's backs. Parsons declared that any new wage law was unlikely, and advised fishing servants not to hire out to planters whose merchants would not secure wages. Fishermen, according to this stalwart Liberal organ, should now shoulder the responsibility for themselves as free players in the marketplace.[58] The split in Liberal ranks led to Morris's failure, in 1843, to introduce another fisheries act in the Assembly.[59]

When Governor Harvey committed the government to a road and agriculture development plan to alleviate the relief problem in 1843, his action deprived his Liberal antagonists of one of the main levers they had been using to pry responsible government out of the British government. As far as the Colonial Office was concerned, Harvey was trying to prove that the financial burden of relief could be minimized without responsible government.[60] The *Patriot*, using the occasion of the second reading of Morris's proposed Fisheries Act, jumped back on the bandwagon of the wage and lien issue as an alternate grievance with which to justify demands for responsible government.[61] Unhappy with Harvey's lack of support for responsible government, and with the manner in which he was winning over many Liberal leaders to support the government,[62] The *Patriot*, formerly a strong supporter of the Carson-Morris demand for agricultural diversification, now condemned the governor for doing just that:

See to it, then, ye who patronize Agricultural Societies and sound the trumpet of Agriculture far and wide! Your *first* attention should be directed to the skiff and not to the plough – for be assured your Agriculture has an unproductive and unsafe substratum if it be not based upon the produce of our teeming coasts. When shall we see the Governor of Newfoundland presiding as the Patron of a Newfoundland Fishermen's Society and a Bill introduced by the Executive to protect the rights and privileges of Fishermen? Not until we shall have introduced among us the boon of *Responsible Government*.[63]

In 1845 the justices of the Supreme Court (of which James Simms was now a member along with a new chief justice, Thomas Norton, and George Lilly) again confirmed that the wage and lien system no longer formed part of statute or common law.[64] Condemning Norton, the *Patriot* associated him with the outrage of Boulton, and again demanded responsible government.[65] When 'A Fisherman' from Carbonear wrote to say that he and his fellows were no longer willing to accept the tyranny of having no wage lien against supply merchants, and would hold public rallies to support Morris's fishery bill, the *Patriot* recommended this course of action.[66] It enthusiastically greeted a subsequent rally on 8 January 1845 at Harbour Grace.[67] When Morris's latest fishery bill met defeat in 1846, the *Patriot* again indicated that its first commitment was to responsible government. Any other issue, whether it be agricultural diversification or the wage lien, was simply a ploy to use in its attainment:

for we have proof enough that under no other system than that of responsible Government, could the fisherman hope for justice. Under the present Executive a new impetus had been given to a novel object – to agriculture – which compared with the substantial interests of the Fisheries, was but a glittering shadow. The governor of the colony from the Throne, had even recommended Bounties and Premiums to be liberally bestowed upon the ploughman, but the Fisherman on whom every other class in the country depended, was disregarded and uncared for. Did not this lamentable state of things show that responsibility was sadly needed in the Councils of the Colony?[68]

Throughout the late 1840s and into the 1850s the debate over the wage and lien system continued. Sectarian struggle, sparked by debates on issues of school funding, increased representation in the assembly, and reciprocity with the Americans, finally led to responsible government in 1855.[69]

Reformers quickly moved to throw a bone to the fishermen whose support they had cultivated on the wage law issue. The new Liberal government, in the fall of 1855, passed an act declaring that fishing servants did have a preferential lien for their wages on fish received from planters by merchants. The working of the wage and lien system had little material effect on the history of planters' development on the northeast coast because the new wage act was symptomatic of the long-term structural problems faced by northeast-coast planters. Just as the earlier 1824 act defined the wage lien in terms of planter insolvency, so too did the 1855 act.[70] The new wage law reflected the strong links which had persisted between the employment of wage labour and planter failure in the northeast coast fishery.

By 1855 planters had found their own solutions to the problem of employing wage labour in the northeast-coast fishery by relying on family labour and the share system. When Chief Justice Boulton arrived in the colony in 1833 few people believed that the wage and lien system actually encouraged capital accumulation or widespread employment opportunities for servants in the fishery. Previous chief justices and government officials had indeed recommended its abolition in an attempt to encourage planter prosperity. Boulton, by not recognizing the existence of the liens for wages and current supply, simply followed the line of his predecessors.

There is no evidence to suggest that the wage and lien system underwrote planter or servant prosperity, just as little evidence exists to prove that merchants undercut it. But the history of the wage and lien system, like that of the merchants' relationship to agriculture, was rewritten by Liberal campaigns for constitutional reform. Liberal leaders, particularly Robert Parsons of the *Patriot*, invented an historical tradition which suggested that the liens for wages and current supply were the custom of the resident fishery. Boulton's decisions served as a convenient 'outrage' in Liberal mythology: the chief justice was a foreign Tory, an arbitrary dupe of the colony's outport fish merchants who was out to stop any challenge to the merchants' monopoly of the fish trade. While this issue may have had only a small place within larger Liberal struggles, it did require the rewriting of the planters'

experiences. Not understanding that the political terrain of Newfoundland demanded that he see the Roman Catholic and servant interest
as one and the same, Boulton lost the paternal, tactical edge which
his fellow Upper Canadian Tories used so well in their general election
of 1836. In Newfoundland the Liberals used the wage law issue to
accommodate the producing classes. However, in doing so they convinced many that before Boulton took office planters prospered
through the employment of servants in the fisheries, only to be undercut by his collusion with greedy merchants.

Conclusion

constructs in
history

Reform and, later, Liberal struggles for constitutional change led to the chimerical reinterpretation of the history of the Newfoundland fishery: that merchants dominated fishing families through truck by refusing to allow residents to develop alternate forms of production or ways of organizing labour that would lessen their dependence on merchant credit. The underpinning of the myth – that merchants prevented domestic capitalist development in the early nineteenth century by opposing economic diversification in agriculture and manipulating the law – is untrue. Merchants, state officials, and fishing families had long realized that Newfoundland's soil and climate could not support petty production in agriculture alone. Although they were not enthusiastic supporters of fishing people's agricultural activities, the credit strategies of the merchants encouraged such activities anyway. Unwilling to risk overextension of capital, especially in the wake of the post-1815 depression, merchants tried to extend as little credit as possible for provisions that fish producers might want to take on account. Merchants had to supply some capital equipment, such as nets and hooks, if fishermen were to provide the staple commodities of their trade, but they could cut back on the amount of food they were willing to advance fishing families on credit. British authorities on the island cooperated with merchants by allowing households to cultivate what land they needed to provide for their winter subsistence.

But gardens + commercial
agric rather than subsistence

Garden cultivation was one area in which fishing households demonstrated that they were not simply tied passively to the domination of merchant capital. But it is always worth keeping in mind that the land the settlers tried to farm was infertile in comparison to Cape Breton, the much more fruitful countryside of the Canadas, or the rest of the Maritimes. Unlike the farmers of the other British North American colonies, fishing people could produce for themselves little of the means by which they could escape dependence on merchants, let alone goods which might have served as alternate industrial inputs or attracted mercantile trade and investment.

However, fishing people were not helpless victims in the paternalistic relations of truck. By cultivating the soil, in an effort to lessen their dependence on merchant credit, northeast-coast households took the lead over government policy and merchant-credit restriction in trying to find ways to deal with the problems of the fishing industry.

Planters and servants, moreover, did not accept merchants' and masters' use of truck without challenge, but often resorted to court action or direct violent confrontation to limit their exploitation by price manipulations of their accounts. Fishing households, furthermore, were not made up of completely impoverished, abject people confronted by wealthy merchants. Court cases involving masters and servants indicate that considerable differences existed among these households. During the unusual market conditions of the Napoleonic era, such differentiation increasingly seemed to promise industrial-capitalist production through some planters' use of wage labour. The promise was not realized, not because conservative merchants opposed it, but because of the post-war collapse of fish prices, the loss to the French of the north-shore fishery, and renewed American competition in the industry.

Little evidence exists to suggest that fishermen's use of hired labour in any form was otherwise persistent or widespread. The fishery of the northeast coast came to rely primarily on the labour of families within households, supplemented by servants usually hired on shares, at those times when the family could not supply enough labour on its own. The offspring of these households sought work as servants in the seal and Labrador fisheries as supplements to family resources, perhaps as a transitional stage on the way to the establishment of their own households. For some in the fishery, differentiation continued, not by hiring wage labour in fish production, but by becoming petty merchants who coasted for trade in the Labrador fishery. Successful planters in the Labrador fishery, like merchants, used truck to minimize wage payments to servants, and supplied freight and supply services to families who made the annual migration to fish on the Labrador coast.

The same legal system which gave fishing people a forum in which to challenge truck, ironically served as the main constraint on increased use of wage labour in fish production. The rigidity of the wage and lien system embedded in Palliser's Act encouraged planters to continue to use labour hired on shares to supplement family-supplied labour. The non-wage labour requirements of the fishing household ensured that even female labour would be closely integrated into staple production for the market.

Women's role in the fishery aggravated the inability of northeast-coast agriculture to generate the surpluses in nonmarket production which provided the basis for so much of the early domestic industry of places like Upper Canada. The crucial work in which fishing women did engage to maintain their households – an underpinning of the staple trade as much as that of their families – was in turn constrained by the region's poor agricultural resources.

Root crops, the main garden produce of the northeast coast, like salt fish, required little processing and created few local linkage effects. Neither root crops nor cod provided the raw material for diversified household manufacture, which might have allowed people to avoid reliance on merchant credit let alone produce commodities for the domestic market. Garden cultivation, especially of potatoes, did allow fishing people to survive post-1815 credit restrictions. The survival of northeast-coast planters tenuously rested on one crop (potatoes), seals, and a cod fishery characterized by intense international competition and partial restriction of access to the Labrador fishery with its poorer-quality product.

Not everyone fought with each other over topsoil, eggs, or potatoes, but the existence of such fights suggests just how difficult it was for fishing people to coax food from the soil. If such difficulties were not bad enough, disaster struck in the 1830s and 1840s when potato rot and bad weather combined almost annually to see the crop fail. Merchant credit did not fill the resulting provision gap, and the fate for many appeared to be famine.

The near-famine years from 1832 to 1855 were hard times for Newfoundland fishing people in more ways than one. As was the case during the provisions crisis of 1816–17, some northeast-coast residents responded collectively to food shortages during the winter of 1832–3 by forcing open merchants' stores. But these were not to be years in which people accepted the futility of placing much hope in Newfoundland's limited agriculture. Fishing people did not engage in any kind of concerted protest designed to secure an improved return for their labour in the fishery. Few openly questioned the problem of merchant credit practices in the provisions crisis, as political culture became dominated by the shibboleth that fishing people's problems stemmed from overdependence on one resource and merchants were responsible for this overdependence by preventing agricultural diversification.

This reform-created 'merchants-against-agriculture' mystification found favour with imperial officials anxious to avoid the expense of government relief. Reformers and Liberals promised that a Newfoundland assembly could encourage economic diversification through proper agricultural development. As part of their effort to build popular support for their cause, reformers portrayed merchants as being responsible for the crisis because they opposed agriculture, not because they restricted or manipulated credit. Reformers wanted to break the oligarchy of imperially appointed officials and merchants, but representative government failed in this objective. Further, governors willingly co-opted reform enthusiasm for agriculture. Some Liberals responded by taking up the cry of 'outrage' over a supposed merchant-official plot to subvert fishermen's 'rights' in the wage and lien system. When government-sponsored agriculture proved not to be viable, officials and newspapers alike blamed the supposed sloth of the people, and told them to eat what they had formerly treated as manure or garbage rather than look to government coffers for relief.

Liberal campaigns, to be sure, identified merchants' use of credit in the fishery as a contributing factor in the problems of the 1830s–40s. But their strategy was to encourage fishing people to believe that a responsible Liberal government would finally sponsor the flowering of Newfoundland agriculture. Liberals further promised that reinstatement of the wage and lien system would force merchants to allow fishermen their just return, mystifying a legal regime, which to the contrary, had reinforced truck and planters' capital-accumulation problems.

Newfoundland political culture foundered on the rocks of the Liberal mythology. The day-to-day struggles between fishing people and merchants over the workings of credit revealed that fishing people were capable of identifying merchant exploitation as a significant problem in their lives. Credit restrictions made people and government alike aware that merchants were not interested in doing anything with the earnings of the fishery that did not benefit their own account books. Finally, popular support for the Liberals in their struggle for responsible government revealed that the anger of many fishing people towards merchants was a potent political force that could be shaped and led.

But the Liberal party did not direct this popular energy towards change in the fishery. We can glean something of the significance of this failure by briefly considering late Newfoundland history. Responsible governments after 1855 continued to embrace the notion that Newfoundland could only improve economically if its economy diversified. Colonial governments did attempt to improve the fishery by regaining the French Shore, regulating American participation in the fishery, and improving access to overseas markets. The imperial government was responsible for treaty negotiations in these areas. While it would consider Newfoundland's interests, the British government would not allow the colony to take unilateral action; and throughout the late nineteenth century, if Newfoundland's interests clashed with those of Canada, the United States, or France, then the British tended to favour these three as their more important partners.[1]

No mere chicanery underlay the Newfoundland merchants and government's turning to the land. Unlike the fishery, landward industries fell under the direct control of the colonial government, over which the Newfoundland bourgeoisie exercised much more influence. Yet the continuing problems of relief and unemployment demanded change, and the bourgeoisie, building on Liberal misconceptions, believed the answers lay on the land. Throughout the 1880s and 1890s the Newfoundland government began to sponsor an intensive diversification plan which one of Newfoundland's most perceptive historians, David Alexander, aptly called the colony's national policy: an attempt to emulate the Canadian experience by railway development, in order to open the interior to settlement and landward resources to extraction.[2] While governments eventually placed their faith in mining and pulp and paper development in the early twentieth century, two of the national policy's early propagandists, Joseph Hatton and Moses Harvey, made it clear that the reform fantasy of agriculture was their inspiration. 'There is no doubt as to the excellence of the soil in the interior' for crops and pasturage, they claimed. 'All that is necessary to their development is the completion of the railway system now in the course of construction.' Settling people in the interior would not only diversify the economy, but would immediately remove them from relief roles.[3]

Newfoundland was not like Canada; the railways did provide significant temporary employment in construction, but few permanent jobs and little industrial diversification. Railway development did burden government with debt, which almost bankrupted the colony in 1895. For some this crisis produced a longing for the good old days implicit in the Liberal mythology about the wage and lien system: if moving forward by way of national policy has failed, go back to the way things were. The problem was that few remembered the past as a time of planter insolvency or retreat into household production amid that old legal regime. The Newfoundland fishery was remembered as more than a source of wealth for merchants, 'for the planter it was little less so, and for the common labourer (the fisherman) it was unequalled by any field of industry ... in the wages which it brought him.'[4] For merchant James Murray the financial collapse of 1895 and the Liberal romanticizing of planter success under the old regime led to a logical question: 'where are all the wealthy planters of Brigus, Bay Roberts, Carbonear, Harbour Grace, and Trinity Bay?' They, argued Murray, had been betrayed by government policies since 1855. The only thing to do was return to a previous time, when planters supposedly flourished by the benevolent providence of merchant credit and government consisted simply of the summary justice administered by Surrogate judges.[5]

More liberal-democratic, if equally misguided, heads prevailed. The Newfoundland economy slowly recovered as war approached. Merchants continued to invest in limited local manufacturing, and government pursued its national policy. Only William Coaker and the Fishermen's Protective Union (FPU) tried to make the fishery the centrepiece of Newfoundland's economic policy. Fooled neither by the prospects for internal diversification or some legendary good old days in the fishery, Coaker tried to organize fishing people into marketing cooperatives to break the merchants' hold on their industry, and the FPU, acting as a political party, elected Coaker and others to the House of Assembly. Coaker unfortunately had to build alliances with other parties in his attempts to secure better regulation of the fishery. His political allies, committed to diversification, had little faith in the fishery. While merchant opposition and a conscription crisis were important factors in the failure of the FPU, so too was a political climate in which the fishery seemed a weak, backward industry which many believed Newfoundland should no longer depend on alone. Even Coaker came to believe this, and he supported the Newfoundland government's efforts to establish a pulp and paper industry on the island's west coast.[6]

Its retreat from Coaker's fisheries regulations suggests that the Newfoundland government saw the fishery as the private concern of merchant capitalists, not the state. But faith in the ability of landward
diversification to provide long-term solutions to Newfoundland's economic problems allowed politicians the luxury of not having to ask
themselves about the wisdom of remaining committed to largely un-
regulated private enterprise in an industry upon which so many people
depended; and governments continued to neglect the fishery even as
they laboured under the mounting debts of continued diversification
policies.

Newfoundland had now fallen almost completely into the hands of
a political generation which accepted the old Liberal creed. Thus, a
young J.R. Smallwood wrote in 1931 that Newfoundland's problems
stemmed from its long underdevelopment by fish merchants, that the
island was saved only by the appearance of the Liberals and responsible
government. He supported the Newfoundland national policy, calling
it the real beginning 'of the modern era of progress – and that means
industrial progress,' an age of 'the tentative beginnings in industrial
development, growing ever stronger and more daring until to-day the
symbol of the national ideal is a hydro-electric power station rather
than a codfish, a great smokestack rather than a seal.'[7]

This history of progress in Newfoundland, Smallwood claimed, was
the progeny of William Carson, 'its author, its inspiration, its architect
and its leader.'[8] While Carson was not the sole perpetrator of delusions
about Newfoundland's capacity for diversification from which people
like Smallwood suffered, the debt which resulted, further worsened
by the expense of war, drove Newfoundland into further depression,
near-bankruptcy, and then into a form of trusteeship in 1934 when
it lost democracy to government by appointed commission.

The Commission Government did try to improve the fishery, but
it, too, pursued further agricultural settlement and industrial diversification schemes because it could not accept Newfoundland's almost-
complete dependence on the fishery. Ultimately, internal dissent and
a lack of financial resources prevented the commission from doing
much of anything. Only the coming of World War II saw the return
of prosperity to Newfoundland, a prosperity built on military-base construction and a revived demand for fish.

Many Newfoundlanders did not want to return to the conditions
of the Great Depression. Mistakenly associating those bad times with
dominion status and responsible government rather than the ruinous

diversification policies of the past, popular support grew, as did British government favour, for Smallwood's confederation campaign. Small-wood promised Newfoundlanders that he would lead them forward into economic progress, but his provincial government returned to the old ways. It fostered landward diversification in an orgy of spending, which produced little but a crippled oil refinery, a few mines (now mostly closed except for those in Labrador), and pulp and paper plants heavily dependent on government assistance, environmental concessions, wage cutbacks, and labour-eliminating technology. The problem was not so much that Smallwood ignored the fishery – every Newfoundland government had some elementary fisheries policies – but that he continued to expound in the most bombastic ways that it was only one industry among a great many possibilities. He was not prepared to accept that Newfoundland was not a cornucopia of natural-resources, or that the fishery was the only industry capable of sustaining much of a resident population.

Confederation, in any event, brought much of the fishery under federal regulation. Canadian mandarins concentrated on 'rationalizing' the fishery, making it more capital intensive and professionalizing its workers, in the mistaken belief that other sectors of the economy would develop to absorb the labour made superfluous by these changes. The inshore fishery had to continue to provide employment for most outport fishing people, while new fishing ships, Canadian and otherwise, scoured the offshore.

Today we have a complete moratorium on fishing because such over-capacity has contributed to the near-destruction of fish stocks, while creating a Newfoundland population heavily dependent on new forms of government relief because government pursuit of the chimera has done little to strengthen the economy. Failed diversification has contributed to an enduring legacy of government debt, which continues to make relief – unemployment insurance, welfare, or fisheries assistance packages – as difficult as ever for authorities to maintain.[9]

It is not possible to blame all of Newfoundland's problems on the Liberal conception of the chimera. Yet throughout most of the nineteenth and twentieth centuries the Newfoundland fisheries have rarely been at the centre of government policies. From the beginning of their residency, fishing people had done what they could on land and sea to build lives for themselves. The fishery alone proved capable of generating considerable wealth, most of which went to the merchants. But after 1815, even as merchants' withdrawal of credit provoked a crisis, almost no clamour demanding a reorganization of the fishery arose. Settlers redoubled their attempts to farm, hurried on by a government anxious to minimize relief expenditures. While many simply left Newfoundland, those who stayed embraced the Liberal recommendation that government worry more about agriculture and the laws of wage and lien. Too few worried about a better administration of the island's only successful industry. Newfoundland's other resources proved unable to replace the fishery adequately as the source of the livelihoods of most of its people.

The chimera, nonetheless, triumphed. Newfoundlanders came to believe that economic diversification like that of the rest of British North America, rather than any change in the class relations that persisted in the fishery, was where their salvation lay.

Appendix A:
The Law of Wage and Lien

Imperial authorities codified the wage and lien system in two pieces of legislation after years of confused practice in the migratory fishery. The first was Palliser's Act (15 Geo. III, c. 31), passed in 1775 as 'An act for the encouragement of the Fisheries carried on from Great Britain, Ireland, and the British Dominions in Europe, and for securing the return of the fishermen, sailors, and others employed in the said fisheries, to the ports thereof, at the end of the fishing season.' The second, the Judicature Act (32 Geo. III, c. 46), appeared in 1792 as 'An act for establishing courts of judicature in the island of Newfoundland, and the islands adjacent.'

While Palliser's Act contained a variety of regulations concerning bounties, custom tariffs, and trade, only Sections 12 through 18 dealt with wages and liens:

12. And whereas it has been a practice of late years for divers persons to seduce the fishermen, sailors, artificers, and others employed in carrying on the fishery, arriving at *Newfoundland*, on board fishing and other vessels from *Great Britain*, and the *British* dominions in *Europe*, to go from thence to the continent of *America*, to the great detriment of the fishery, and the naval force of this kingdom: Now, in order to remedy the said evil, and to secure the return of the said fishermen, sailors, artificers, and others employed as aforesaid, to the *British* dominions in *Europe*, be it further enacted by the authority aforesaid, That, from and after the first day of *January* one thousand seven hundred and seventy-six, it shall not be lawful for the master, or person having the charge or command of any ship or vessel trading to or from any place within the government of *Newfoundland*, to carry or convey, as passengers, any such fishermen, sailors, artificers, and others, employed as aforesaid, from

thence to any part of the continent of *America*, without the permission under the hand and seal of the governor of the said island of *Newfoundland*, under the penalty of forfeiting two hundred pounds for every such offence.

13.   And whereas in several acts, passed in the eleventh and twelfth years of *William* the third, the eighth of *George* the first, and second and twelfth of *George* the second, provision has been made to prevent seamen and mariners in the merchant service being wilfully left beyond sea, and to secure and provide for their return home to such part of his Majesty's dominions whereto they belong: and whereas, for want of such provisions being extended to seamen and fishermen going out as passengers to *Newfoundland*, and hired and employed in the fisheries carried on there, great numbers of them remain in that country at the end of every fishing season, who would otherwise return home, and some of them have frequently turned robbers and pirates; for remedy of which evil, be it enacted by the authority aforesaid, That no person or persons whatsoever shall, from and after the first day of *January* one thousand seven hundred and seventy-six, employ, or cause to be employed at *Newfoundland*, for the purpose of carrying on the fishery there, any seaman or fisherman going as passengers, or any seaman or fisherman hired there, without first entering into an agreement or contract in writing with every such seaman or fisherman, declaring what wages such seaman or fisherman is to have, and the time for which he shall serve, which shall be signed by both parties; wherein it shall be stipulated (amongst other things) that the person so hiring or employing shall be at liberty to reserve, retain, and deduct, and he is hereby authorised, required, and directed to reserve, retain, and deduct, out of the wages of every person so hired or employed, a sum of money equal to the then current price of a man's passage home, not exceeding forty shillings for each man, which money such hirer or employer shall, at the end of each fishing season, or at the expiration of the covenanted time of service of such seaman or fisherman, pay, or cause to be paid, to the master of a passage or other ship, who shall undertake or agree to carry such seaman or fisherman to and on board such passage or other ship, taking the master's receipt for the passage money, which receipt he shall immediately thereupon deliver to such seaman or fisherman.

14.   And be it further enacted by the authority aforesaid, That no hirer or employer of any such seaman or fisherman shall pay or advance, or cause to be paid or advanced, to such seaman or fisherman, in money, liquor, and goods, or either of them, during the time he shall be in his service, more than one half of the wages which shall at any time be due to him; but such

hirer or employer shall, and is hereby required and directed, immediately at or upon the expiration of every such man's covenanted time or service to pay either in money, or in good bills of exchange, payable either in *Great Britain* or *Ireland*, or in the country to which such seaman or fisherman belongs, the full balance of his wages, except the money herein-before directed to be retained for his passage home; and it shall not be lawful for any such hirer or employer to turn away or discharge any such seaman or fisherman, except for wilful neglect of duty, or other sufficient cause, before the expiration of his covenanted time of service; and in case the hirer or employer of any such seaman or fisherman shall refuse or neglect to comply with any of the terms herein-before mentioned, or shall otherwise offend against this act, every such person so offending shall forfeit and pay, besides the balance that shall be due to such seaman or fisherman, the money herein-before directed to be retained for his passage home, the sum of ten pounds, to the use of such person or persons who shall inform or sue for the same.

15.   And be it further enacted by the authority aforesaid, That in all cases where disputes shall arise concerning the wages of any such seaman or fisherman, the hirer or employer shall be obliged to produce the contract or agreement in writing, herein-before directed to be entered into with every such seaman or fisherman.

16.   And be it further enacted by the authority aforesaid, That all the fish and oil which shall be taken and made by the person or persons who shall hire or employ such seaman or fisherman shall be subject and liable, in the first place, to the payment of the wages of every such seaman or fisherman.

17.   And be it further enacted by the authority aforesaid, That in case any such seaman or fisherman shall at any time wilfully absent himself from his duty or employ, without the leave and consent of his hirer or employer, or shall wilfully neglect or refuse to work according to the true intent and meaning of such contract or agreement, he shall, for every day he shall so absent himself, or neglect or refuse to work as aforesaid, forfeit two days pay to such hirer or employer; and if any such seaman or fisherman shall wilfully absent himself from his said duty or employ for the space of five days without such leave as aforesaid, he shall be deemed a deserter, and shall forfeit to such hirer or employer all such wages as shall at the time of such desertion be due to him, (except so much as is herein-before directed to be reserved and retained for the purpose of paying his passage home); and it shall and may be lawful to and for the governor of *Newfoundland*, or his surrogates, or

the commissary of the vice admiralty court for the time being, or for any justice of the peace in *Newfoundland*, to issue his or their warrant or warrants to apprehend such deserter, and on the oath of one or more credible witness or witnesses to commit him to prison, there to remain until the next court of session which shall be holden in pursuance of the commission of the said governor for the time being; and if found guilty of the said offence at such session, it shall and may be lawful to and for the said court of session, to order such deserter to be publickly whipped as a vagrant, and afterwards to be put on board a passage ship, in order to his being conveyed back to the country whereto he belongs.

18.   And be it further enacted by the authority aforesaid, That all disputes which shall arise concerning the wages of every or any such seaman or fisherman, and all offences which shall be committed by every hirer or employer of such seaman or fisherman, against this act, shall and may be enquired into, heard, and determined, and the penalties and forfeitures thereby incurred shall and may be recovered in the court of session herein-before mentioned, or in the court of vice admiralty having jurisdiction in the said island of *Newfoundland*.

The Judicature Act of 1792 incorporated Section 16 of Palliser's Act into Section 7 concerning the courts' distribution of the effects of insolvent planters. The 1792 act did not limit servants' liens to only fish and oil they actually caught as specified in Palliser's Act:

7.   And be it further enacted, That in the distribution to be made of the estate and effects of such person so declared insolvent, every fisherman and seaman employed in the fishery, who shall be a creditor for wages become due in the then current season, shall first be paid twenty shillings in the pound, so far as the effects will go; and in the next place, every person who shall be a creditor for supplies furnished in the current season, shall be paid twenty shillings in the pound; and lastly, the said creditors for supplies furnished in the then current season, and all other creditors whatsoever, shall be paid equally in proportion, as far as the effects will go, provided that the said creditors for supplies furnished in the then current season shall not be paid more than twenty shillings in the pound on the whole of their debt.

Appendix B:
Selection of Court Record Evidence

The bulk of the evidence used in this book is drawn from the records of the various courts which sat at Harbour Grace from 1785 to 1855 and are now partially preserved at the provincial archives of Newfoundland and Labrador. Two courts dominated Conception Bay's civil suits: the Surrogate Court (GN5/1/B/1) from 1785 to 1825, and the Northern Circuit Court from 1826 to 1855. All of the surviving minutes of the Surrogate Court were read for this book. Cases drawn from this court are cited by box number, minute book date, then date of case. Folio numbers have not been used because damage to the minute books' edges have not made them consistently available. Researchers can find cases cited here in two ways. First, one can simply scan all the cases cited on a particular day until the plaintiff's and defendant's names are found. Second, if folio numbers exist, often an index of cases may be found at the back of a minute book which may be used to direct the researcher to the folio numbers of the case under examination.

The voluminous official minutes of cases heard by the Northern Circuit Court (GN5/2/B/1), and its inferior body, the Court of Session (GN5/4/B/1), contain little more than the names of adversaries, the titles of their disputes, and a brief notation of sentences. No court transcripts are provided. Research, in consequence, was directed towards another set of records which contain evidence for the Northern Circuit Court and the Courts of Session. The Harbour Grace Court Records Collection (GN5/3/B/19) contains two basic types of files: writs issued by the Courts of Session and the Northern District Court by the latter's authority; and miscellaneous documents, trial transcripts, evidence, and judicial correspondence for the years 1825 to

1855. All files for the period 1826 to 1855 were examined. The Harbour Grace court records provide a much fuller array of material on life and labour on the northeast coast than do the circuit courts' official minutes.

The nature of the Harbour Grace court records demanded that some basic sampling be used. The collection consists of seventy-five large archive boxes, each containing approximately ten to eleven files. I read all files containing miscellaneous documents, sampling only writs. I employed no random selection process, because I make very few claims based on quantification. The Provincial Archives of Newfoundland and Labrador collected this material from the old court house at Harbour Grace. Staff picked up material from the floor and stuffed it into file folders without any further organization, ensuring a certain amount of randomness. No finding aids or computerized access to this collection exists. I simply read every fourth writ because they were in very fragile condition, did not appear able to withstand much handling, and generally recorded fairly routine and not very revealing information. Occasionally, however, a court official would have recorded personal observations on the conduct of participants in a particular dispute, and I wanted to be sure that I caught some of these in my research. Sampled writs are identified in endnotes by a writ number at the end of each citation. The following index to selected writs is provided for further reference:

Box 18	File 1	(1829)	6 writs	File 2	(1829)	9 writs
	File 3	(1829)	7 "	File 4	(1833)	7 "
	File 5	(1833)	6 "	File 6	(1833)	10 "
	File 7	(1834)	9 "	File 8	(1834)	6 "
	File 9	(1836)	8 "	File 10	(1837)	7 "
	File 11	(1837)	3 "	File 12	(1837)	2 "
Box 19	File 2	(1829)	10 "	File 6	(1829)	13 "
Box 20	File 1	(1833)	5 "	File 6	(1853)	5 "
	File 9	(1832)	9 "	File 10	(1832)	13 "
Box 21	File 3	(1838)	4 "	File 4	(1838)	5 "
	File 5	(1848)	3 "	File 6	(1848)	3 "
Box 22	File 6	(1832)	8 "	File 7	(1832)	8 "
	File 8	(1832)	8 "			
Box 26	File 1	(1848)	4 "	File 2	(1848)	1 "
	File 3	(1855)	4 "	File 4	(1855)	5 "
	File 10	(1827)	6 "	File 14	(1845)	2 "
Box 27	File 4	(1840)	4 "	File 5	(1840)	4 "
Box 28	File 3	(1837)	9 "	File 5	(1837)	8 "
	File 10	(1837)	3 "	File 11	(1837)	6 "

Box 30	File 1	(1826)	12	"	File 2	(1826)	12	"
	File 3	(1826)	12	"	File 4	(1827)	1	"
	File 5	(1827)	8	"	File 6	(1827)	11	"
	File 7	(1830)	7	"	File 8	(1830)	5	"
Box 34	File 1	(1840)	8	"				
Box 38	File 1	(1852)	3	"	File 2	(1852)	1	"
	File 4	(1830)	10	"	File 5	(1830)	8	"
	File 6	(1830)	5	"	File 9	(1844)	3	"
	File 11	(1844)	4	"				
Box 41	File 2	(1826)	9	"	File 3	(1826)	13	"
	File 4	(1834)	13	"	File 5	(1839)	6	"
Box 42	File 1	(1840)	5	"	File 2	(1840)	6	"
Box 47	File 2	(1846)	2	"	File 3	(1839)	8	"
Box 49	File 3	(1840)	3	"	File 4	(1840)	5	"
	File 5	(1845)	3	"	File 6	(1845)	7	"
Box 50	File 1	(1835)	7	"				
Box 54	File 4	(1831)	8	"	File 5	(1831)	9	"
	File 6	(1839)	6	"	File 8	(1843)	1	"
	File 12	(1839)	6	"				
Box 55	File 2	(1830-5)	2 writs					
Box 56	File 2	(1830)	7	"	File 3	(1830-5)	7	"
	File 4	(1830-5)	5	"	File 5	(1846-7)	2	"
Box 74	File 1	(1842)	6	"	File 2	(1842)	5	"
	File 3	(1842)	2	"	File 4	(1843)	5	"
	File 5	(1841)	5	"	File 6	(1842)	6	"
	File 7	(1841-2)	4	"	File 8	(1849)	1	"
	File 9	(1843)	2	"				
Box 75	File 4	(1841)	6	"	File 5	(1841)	2	"
	File 6	(1841)	2	"	File 7	(1835)	5	"
	File 8	(1842)	4	"				

Total writs sampled: 542 (no count of the total number of writs in this collection exists, but there are well over 2,000).

The sampled writs do not always indicate who won or lost a case. They do provide a wealth of material, incidental to the case, about social and productive relationships in the Conception Bay fisheries. Evidence drawn from the sampled writs, as well as from other documents in the Harbour Grace collection, and those few cases from the Surrogate Court minutes where trial transcripts were actually transcribed and preserved, have been used to supplement government correspondence, newspapers, and missionary correspondence in order to illustrate interpretations offered about the development of northeast-coast society. Evidence was drawn from a further collection, the pre-1826 Court of Session (GN5/4/B/1), to supplement Surrogate Court evidence. The Court of Session heard criminal and petty civil cases when the Surrogate Court was not in session.

Writ samples were broken down into basic categories for descriptive purposes (see table 7). Sixty-eight per cent of the writs involved simple debt disputes covering everything from payment defaults on promissory notes and merchants' suits for account payments to a variety of petty-debt disputes between residents of Conception Bay. This book focuses on wage and insolvency disputes because historiography identifies these as having a particular importance in capital accumulation on the northeast coast.

TABLE 7
Sample of writs from the Northern Circuit Court, 1826–55

Year	Wages	Debt	Insolvency		Land/lease	Other	Total
			Lab.	Inshore			
1826	6	46	0	0	1	5	58
1827	3	16	0	1	3	3	26
1828	-	-	-	-	-	-	-
1829	7	29	0	0	1	7	44
1830	3	36	0	0	9	6	54
1831	2	12	0	0	1	2	17
1832	7	37	0	0	2	7	53
1833	7	12	1	0	4	6	30
1834	2	21	1	0	2	2	28
1835	2	9	0	0	0	1	12
1836	0	7	1	1	0	0	9
1837	5	22	2	1	5	3	38
1838	0	7	0	0	0	2	9
1839	0	21	0	0	1	4	26
1840	6	24	0	1	0	4	35
1841	1	15	0	0	2	1	19
1842	5	14	0	0	3	3	25
1843	0	5	1	0	0	0	6
1844	4	1	0	0	0	2	7
1845	1	10	1	0	0	0	12
1846	2	1	0	0	0	0	3
1847	0	0	0	0	0	0	0
1848	1	9	0	1	0	1	12
1849	0	1	0	0	0	0	1
1850	-	-	-	-	-	-	-
1851	-	-	-	-	-	-	-
1852	0	3	0	0	0	1	4
1853	2	3	0	0	0	0	5
1854	-	-	-	-	-	-	-
1855	2	5	0	1	1	0	9
Total	68	366	7	6	35	60	542

Notes

(form here – authored does not specify PhD, Book or article)

PREFACE

1 An introduction to the Newfoundland historiography can be found in Peter Neary, 'The Writing of Newfoundland History: An Introductory Survey,' in Hiller and Neary, *Newfoundland*, 3–15; and in Keith Matthews, 'Historical Fence Building,' 21–9.

2 Sider, *Culture and Class*, 7–37, 112–18; Antler, 'Colonial Exploitation and Economic Stagnation,' 1975. For a more extensive treatment of the latter historiography's relationship to the earlier Newfoundland literature, see Sean T. Cadigan, 'Planters, Households and Merchant Capitalism: Northeast-Coast Newfoundland, 1800–1855,' in Sansom, *Contested Countryside*, 150–79.

3 Dobb, *Development of Capitalism*, 17–127. The ensuing debate has been reprinted in Hilton, *Feudalism to Capitalism*. The next phase of the transition debate can be found in Aston and Philpin, *Brenner Debate*. For the major currents in dependency theory and world systems analyses, see Frank, *World Accumulation* and *Dependent Accumulation*; Wallerstein, *Historical Capitalism* and *Capitalist World-Economy*. For criticisms of this latter literature, see Brenner, 'Origins of Capitalist Development,' and Fox-Genovese and Genovese, *Merchant Capital*.

4 Among the more prominent works marked by aspects of this perspective, in chronological order of appearance, are Innis, *Cod Fisheries*; Keith Matthews, 'England-Newfoundland Fishery'; Head, *Eighteenth Century Newfoundland*; Ryan, *Fish Out of Water*; Handcock, *there comes noe women*.

5 DuPlessis, 'World Systems,' 20–6; Mendels, 'Proto-industrialization,' 241–61; Medick, 'Proto-industrial Family Economy,' 291–316; and Clark, 'Household Economy,' 174–5.

6 Ommer, 'Introduction,' in Ommer, *Merchant Credit*, 10.
7 Palmer, 'Social Formation and Class Formation in North America, 1800–1900,' in Levine, *Proletarianization and Family History*, 230–43; and Kulikoff, *Agrarian Origins*, 25–32.
8 See McCalla, *Planting the Province*.
9 McNally, 'Commodity Fetishism,' 35–63; 'Technological Determinism,' 161–70; and 'Political Economy,' 217–26.
10 Ommer, *Outpost to Outport*.
11 R.E. Baldwin, 'Patterns of Development,' 161–79.
12 Cadigan, 'Staple Model Reconsidered,' 48–71.
13 Head, *Eighteenth Century Newfoundland*, 223; Story, Kirwin, and Widdowson, *Dictionary of Newfoundland English*, 187, 326–7, 381–3, 399, 461–63, 525; Keith Matthews, *Lectures*, 177; and Handcock, *there comes noe women*, 27, 106.

INTRODUCTION

1 Bayly, *Imperial Meridian*, 76–94, 157. For the sake of brevity I have simplified the Maritime experience, especially that of Cape Breton. See MacNutt, *Atlantic Provinces*, 76–102; Craig, *Upper Canada*, 1–41; Brian Young and John A. Dickinson, *Short History of Quebec*, 59–64; and Neil MacKinnon, *This Unfriendly Soil*, 16–26, 67–88. Much of my thought about the development of colonial government has been shaped by my earlier research on Upper Canada. See Cadigan, 'Paternalism and Politics,' 319–47.
2 Akenson, *Irish in Ontario*, 138–63.
3 McCalla, *Planting the Province*, 10.
4 In addition to McCalla's *Planting the Province*, Akenson's *Irish in Ontario*, and subsequent specific citations, the following interpretation of the Upper Canadian experience is based on my general understanding of the following literature: Acheson, 'York Commerce,' 406–28; Peter Baskerville, 'The Entrepreneur and the Metropolitan Impulse: James Grey Bethune and Cobourg 1825–36,' in Petryshyn, Calman, Crossen, and Dzubak, *Victorian Cobourg*, 11–66; Gilmour, *Manufacturing in Southern Ontario 1851–1891*; Johnson, *Ontario 1615–1875*; McCalla, *Upper Canada Trade*; Gagan, *Hopeful Travellers*; and Brian S. Osborne, 'Trading on a Frontier: The Function of Peddlers, Markets, and Fairs in Nineteenth-Century Ontario,' in Akenson, *Canadian Papers*, vol. 2, 59–82.
5 This is a brief summary of *Enterprises of Robert Hamilton*, by Bruce G. Wilson.
6 Cohen, *Women's Work*, 6–12.
7 For purposes of argument I have neglected the considerable ethnic and

legal issues which arise in the Lower Canadian case to emphasize the essential agrarian nature of its development to the mid-nineteenth century. My interpretation of the Lower Canadian agricultural experience is based on F. Lewis and R.M. McInnis, 'French-Canadian Farmer,' 497–514; R.M. McInnis, 'A Reconsideration of the State of Agriculture in Lower Canada in the First Half of the Nineteenth Century,' in Akenson, *Rural History*, vol. 2, 9–49; Serge Courville, 'Villages and Agriculture in the Seigneuries of Lower Canada: Conditions of a Comprehensive Study of Rural Quebec in the First Half of the Nineteenth Century,' in Akenson, *Rural History*, vol. 5, 121–49; Greer, *Peasant, Lord, and Merchant*; Greer, *Patriots and the People*, 21–8; Young and Dickinson, *Short History of Quebec*, 103–38; and McCallum, *Unequal Beginnings*.

8 J.M. Bumsted, '1763–1783: Resettlement and Rebellion,' and Graeme Wynn, '1800–1810: Turning the Century,' in Buckner and Reid, *Atlantic Region to Confederation*, 153–83, 211–33.

9 Robert A. MacKinnon and Ronald H. Walder, 'Agriculture in Atlantic Canada, 1851,' in Gentilcore, *Historical Atlas*, vol. 2, plate 12.

10 Douglas Baldwin, *Land of the Red Soil*, 13, 65–8, 102–3.

11 Wynn, *Timber Colony*.

12 Beatrice Craig, 'Agriculture in a Pioneer Region: The Upper St John River Valley in the First Half of the 19th Century,' in Inwood, *Farm, Factory and Fortune*, 17–36; and T.W. Acheson, 'New Brunswick Agriculture,' 5–26.

13 Alan R. MacNeil, 'The Acadian Legacy and Agricultural Development in Nova Scotia, 1760–1861,' in Inwood, *Farm, Factory and Fortune*, 1–16.

14 Hornsby, *Nineteenth-Century Cape Breton*, 23–84; and Rusty Bittermann, 'Hierarchy of the Soil', 33–55; and 'Farm Households and Wage Labour in the Northeastern Maritimes in the Early 19th Century,' *Labour/Le Travail*, 31 (1993), 13–45.

15 R.C.B. Risk, 'The Law and the Economy in Mid-Nineteenth Century Ontario: A Perspective,' in Flaherty, *Essays*, 91–112.

16 Paul Craven, 'The Law of Master and Servant in Mid-Nineteenth-Century Ontario,' in Flaherty, *Essays*, 175–211; and Palmer, *Working-Class Experience*, 65.

17 Akenson, *Irish in Ontario*, 139–201.

18 In addition to my own 'Election of 1836,' my interpretation here is based on Read and Stagg, *Rebellion of 1837*, xxi–xvi; Greer, *Patriots and the People*, 88–118, 286–93; and Palmer, *Working-Class Experience*, 69–75.

19 On the differences between the Canadas and the Maritimes, see Buckner, *Responsible Government*, 8–80. On Nova Scotia, see Beck, *Joseph Howe*, 177–8; On New Brunswick, see W.S. MacNutt, *New Brunswick, A History:*

1784–1867 (Toronto: Macmillan 1963), 278–345. See also T.W. Acheson, 'The 1840s: Decade of Tribulation,' in Buckner and Reid, *Atlantic Region*, 326.

20 Bitterman, 'Escheat!'

21 On Vancouver Island, see Loo, *Law, Order, and Authority*, especially chap. 2.

22 In addition to MacKinnon and Walter, plate 12, see Jean-Claude Robert, Norman Séguin, and Serge Courville, 'An Established Agriculture: Lower Canada to 1851,' plate 13; and J. David Wood, Peter Ennals, and Thomas F. McIlwraith, 'A New Agriculture: Upper Canada to 1851,' plate 14, in Gentilcore, *Historical Atlas*, vol. 2.

23 Acheson, 'New Brunswick Agriculture,' 5.

24 The remarkable paucity of merchant records, and the absence of records generated by the fishing people themselves, meant that I relied heavily on court records, government correspondence, and newspapers for this section. While such records allowed me to infer a great deal about the broad patterns of class differentiation in Newfoundland fishing society, they did not allow a detailed social and demographic analysis of the relationships between merchants, planters, and servants.

CHAPTER 1 Political Economy of the Resident Fishery

1 On Conception Bay as an urban environment, see Sider, *Culture and Class*, 20–1, 89–91.

2 Gerald S. Graham, 'Fisheries and Sea Power,' in Rawlyk, *Historical Essays*, 7–13.

3 Keith Matthews, 'England-Newfoundland Fishery,' 1–10; E.F.J. Mathews, 'Economic History of Poole,' 22–71.

4 Matthews, 'England-Newfoundland Fishery,' 14–21.

5 Gillian T. Cell, *English Enterprise*, 53–95.

6 Handcock, *there comes noe women*, 23–44, 73–84.

7 Ibid., 91–120.

8 Michael Staveley, 'Population Dynamics in Newfoundland: The Regional Patterns,' in Mannion, *Peopling of Newfoundland*, 67.

9 Alan G. Macpherson, 'A Modal Sequence in the Peopling of Central Bonavista Bay, 1676–1857,' in Mannion, *Peopling of Newfoundland*, 112–28.

10 Crabb, 'Agriculture in Newfoundland,' 41–51.

11 Head, *Eighteenth Century Newfoundland*, 13–14, 90–4.

12 Shannon Ryan, 'Fishery to Colony: A Newfoundland Watershed, 1793–1815, in Buckner and Frank, *Acadiensis Reader*, vol. 1, 134–48.

13 Keith Matthews, *Lectures*, 153–5; Little, 'Plebeian Collective Action,' 7–67.

14 Gosse, *Philip Henry Gosse*, 49.

15 Ibid., 46–51.

16 Ryan, 'Fishery to Colony,' 148.

17 Sanger, 'Newfoundland Seal Fishery,' 12-53.

18 Ryan, 'Newfoundland Cod Fishery,' 17-49.

19 Handcock, *there comes noe women*, 91-106. Handcock derived his data from calculations of ten-year means of highly unreliable population statistics collected by Newfoundland governors for the Colonial Office, and discusses the problem of using these in his 'English Migration to Newfoundland,' in Mannion, *Peopling of Newfoundland*, 19-20.

20 Government of Newfoundland, *Population Returns, 1836*, 2-8; *Abstract Census* (1845), 2-13; *Abstract Census* (1857), 13-63.

21 Keith Matthews, 'England-Newfoundland Fishery,' 10-12, 65, 132-7, 156-324.

22 Keith Matthews, *Lectures*, 94.

23 Keith Matthews, 'England-Newfoundland Fishery,' 232-383.

24 Keith Matthews, *Lectures*, 96-102.

25 Reeves, *Government of Newfoundland*, 78, 136.

26 Keith Matthews, 'England-Newfoundland Fishery,' 381-453.

27 Provincial Archives of Newfoundland and Labrador, (hereafter cited as PANL), P1/5, Gov. Thomas Duckworth Papers, Microfilm, reel M-3176, reference copy of 15 Geo. III, c. 31, 547-75.

28 English, 'Newfoundland Legal System,' 89-119.

29 O'Flaherty, 'Carson, William,' 151-6.

30 Mannion, 'Morris, Patrick,' 626-9.

31 Leslie Harris, 'Representative Government,' 46-50; Thompson, 'Cochrane, Sir Thomas John,' 178-80.

32 Gunn, *Political History of Newfoundland*, 3-128.

CHAPTER 2 Fishing Households and Family Labour

1 The association of youth and service in the development of Newfoundland settlement is examined in Handcock, *there comes noe women*, chaps. 1-5.

2 Centre for Newfoundland Studies, Memorial University of Newfound- ⟩ land, Colonial Office Papers No. 194 (hereafter cited as CO 194), vol. 138, 1788-91, Microfilm box 678, fols. 290-2; Chief Justice Reeve's report to Secretary Dundas on the legislation and judiciary of Newfoundland, n.d., ca., 1791.

3 Ryan, 'Fishery to Colony,' in Buckner and Frank, *Acadiensis Reader*, vol. 1, 130-48.

4 CO 194, vol. 40, 1798, box 679, fols. 135-7, Gov. W. Waldegrave to the Duke of Portland, *Agincourt* at sea, 30 Oct. 1798.

5 Keith Matthews, 'England-Newfoundland Fishery,' 593-6.

6 CO 194, vol. 44, 1804–5, box 680, fols. 50–3; Gov. E. Gower to Earl Camden, Hermitage, 24 Dec. 1804; box 681, fol. 141; Gower to Camden, St John's, 18 July 1805; vol. 45, 1806, box 681, fol. 15; Gower to W. Windham, Hermitage, 13 Feb. 1806.

7 Ryan, *Fish Out of Water*, 46–51.

8 CO 194, vol. 45, 1806, box 681, fol. 20; Gov. Gower, 'Explanatory Observations on the Accompanying Return of the Fishery and Inhabitants of Newfoundland,' 1804.

9 PANL P1/5, Duckworth Papers, Microfilm, box 351.5, M-3717, fols. 2762–3, petition of the merchants of Conception Bay to Gov. T. Duckworth, 31 July 1812.

10 PANL, GN5/1/B/1, Minutes of the Surrogate Court (hereafter cited as Surrogate's Minutes), box 1, 1807–10, Richard Kain v Francis Pike, 8 Feb. 1808. Readers should note that full archival numbering of legal records is used throughout this book, instead of the more usual practice of citing cases by plaintiff and defendant, because of common discrepancies in spelling and detail for individual cases variously reported in official minutes, later printed versions, and ancillary documentary collections.

11 Ibid., Michael Kain v William Peddle, 11 Jan. 1808.

12 Ibid., Robert Ash v Elizabeth Pike, 2 June 1808.

13 For other court cases which reveal that planters hired wage labour to catch fish both on the north shore and at Labrador, as well as subcontractors to cure the fish (see note 10 above). Box 1, Surrogate's Minutes, 1807–10, Hunt & Co. v James Quinlan, 18 Dec. 1809; Surrogate's Minutes, 1813–15, Michael Power v James Rafter, 12 May 1814, box 2, 1816–24, Surrogate's Minutes, 1816–18; Michael Purcell v William Donovan, 22 May 1817; Macfarlane & Scott v James Fox, 19 Dec. 1817; Minutes, 1818–19; Colbert and Driscoll v Henry Webber & Co., 28 Jan. 1819.

14 CO 194, vol. 55, 1814, box 685, fol. 95; Gov. R.G. Keats to Earl Bathurst, *Bellerophon* at sea, 29 Dec. 1814.

15 Ibid., vol. 59, 1817, box 686, fols. 110–12; Gov. Pickmore to Henry Goulbourn, Portsmouth, 20 July 1817. Petitions came from the merchants of Bristol, mainly trading to Conception Bay: Thomas Thorne & Co., William Danson, William Henderson, William Mullowney, and Bartholomew Henderson & Co.; and the merchants of Poole, trading to all parts of the northeast coast: Thomas Colborne, George Garland, Chris Spurrier & Co., Samuel and Jn Clark, John Slade & Co., Robert Slade, Slade & Cox, Pack, Gosse & Fryer, George & J. Kemp & Co., Sleat and Read, and Joseph Bird. Also vol. 60, 1817, box 687, fols. 185–6, 211–12; petitions of the merchants of Poole and Bristol, 7, 13 March 1817.

16 Ibid., vol. 60, 1817, box 687, fols. 249–75; J. Newart to Earl Bathurst, St John's, 4 March 1817. Newart's observations were confirmed in the British House of Commons Select Committee testimony of George Garland and George Kemp Sr and Jr, merchants of Poole, and J.H. Attwood, merchant of St John's. Ibid., fols. 290–312; 'Report from the Select Committee on the Newfoundland Trade; with Minutes of Evidence,' 26 June 1817.

17 PANL, Wesleyan Methodist Missionary Society, Newfoundland Correspondence (hereafter cited as WMMS Corr.), 1818–24, microfilm box 971.8W2, drawer B-4-4, no. 68; John Walsh to Joseph Taylor, Conception Bay, 10 July 1819.

18 Ibid., no. 137; 'Observations &c. on the island of Newfoundland,' March 1819.

19 John Mannion and W. Gordon Handcock, 'Origins of the Newfoundland Population, 1836,' in Gentilcore, *Historical Atlas*, vol. 2.

20 CO 194, vol. 63, 1820, box 688, fols. 250–67; Capt. J. Nicholas to Lord Melville, Newfoundland, 18 Oct. 1820.

21 Ibid., vol. 64, 1821, box 688, fols. 139–40; grand jury to Forbes, St John's, 13 Oct. 1821.

22 CO 194, vol. 72, 1826, box 693, fols. 87–100; Gov. T. Cochrane to Earl Bathurst, St John's, 30 Jan. 1826.

23 PANL, WMMS, Corr., 1818–24, 971.8W2, B-4-4, no. 173, 'Observations &c. on the island of Newfoundland,' March 1819; ibid., 1824–5, 971.8W5, B-4-4, no. 171; William Wilson to the Wesleyan Methodist Missionary Committee; Port de Grave, 6 Sept. 1825.

24 PANL, GN5/1/B/1, Surrogate's Minutes, Box 2, 1816–18, petition of William Taylor, Harbour Grace, 10 June 1817. PANL, GN5/3/B/19, Harbour Grace Court Records (hereafter cited as HGCR), box 30, file 6, writ issued in the Harbour Grace Northern Circuit Court (hereafter cited as HGNCC), no. 131, 21 Nov. 1827.

25 CO 194, vol. 74, 1827, box 695, fols. 167–72; Gov. Cochrane to Viscount Goderich, St John's, 26 Sept. 1827; ibid., vol. 76, 1828, box 696, fols. 324–7; petition of the St John's Chamber of Commerce to Sir G. Murray, 1828; ibid., vol. 87, 1834, box 536, fols. 174–8; petition of the House of Assembly to G.C. Stanley, St John's, 22 May 1834.

26 Ibid., box 536, vol. 87, 1834, fols. 34–5; Gov. Cochrane to E.G. Stanley, St John's, 1 Feb. 1834.

27 PANL, GN5/3/B/19, HGCR, box 33, file 1, 1835–43; ibid., file 2, 1835–43; the Queen v John Sparks and others, 23 May 1840.

28 Ibid., box 22, file 5; Thomas Danson to Colonial Secretary James Crowdy, Harbour Grace, 22 Nov. 1833.

29 Ibid., box 50, file 1, 1821–47; Martin and Jacob v John Long, 14 May 1833 in HGNCC. Writ no. 63; box 18, file 8, 1834; Susan Watts v Richard Taylor, Carbonear, 18 Nov. 1834 in HGNCC. Writ no. 346; box 28, file 5, 1837; Laurence Shea v John Shea, 13 Nov. 1837. Writ no. illegible; box 18, file 12, 1837; Thos. Ridley & Co. v William Thistle, 23 Dec. 1836 in HGNCC. Writ no. illegible.

  For other cases in which servants' suits for wages occasioned their planters' insolvencies in the Labrador fishery, see box 49, file 5, 1845; John Rourke v Edmund Guinea, 29 Oct. 1845, HGNCC. Writ no. 40, box 64, file 2, 1830–9; statement of the debts and effects of Wm. Ash, Harbour Grace, 13 Nov. 1832, box 41, file 6, 1839; Cornelius Byrne v Edward Pike, Carbonear, HGNCC, 6 Nov. 1839, writ no. 123, box 58, file 3, 1840–9; statement of the affairs of William Flynn Jr, 11 Nov. 1844, box 55, file 3, 1840–8; John Munn v Kennedy Thomey, HGNCC, 16 Nov. 1847, box 20, file 5, 1833; 'The petition of John Delaney of Port de Grave, fisherman,' HGNCC, 6 Nov. 1850, box 26, file 10, 1827; petition of Thomas Powell, Carbonear, to Chief Justice R.A. Tucker, 18 May 1827, box 38, file 3, 1830; insolvent estate of John P. Taylor, 2 June 1830 (Taylor owed £600 to Slade, Elson & Co., £300 to Wm. Taylor & Co., £7 to Robinson & Brooking, and £2 to Tocque and Levi), box 20, file 1, 1833, insolvency of Denis Thomey, 21 Nov. 1833, box 20, file 1, 1833, 'property belonging to Edmund Barrett,' box 20, file 1, 1833; account of James Quiddihy with George Forward, 1833, box 18, file 7, 1834, no. 51, writ of HGNCC William Ash Jr and John Taylor to pay 2 pounds to Robert Pack and William Bennett, trustees of Taylor's estate, 1 July 1834, box 18, file 18; petition of Michael Keefe, planter, to E.B. Brenton, judge of the HGNCC, 19 June 1834, box 28, file 4, 1837; Joseph Soper v James Cooney, 6 Nov. 1837, and 'Schedule of the Estate of James Cooney Insolvent,' 20 Nov. 1837.

30 Ibid., box 18, file 12, 1837; estate of Simon Levi, 1837; also box 28, file 10, 1837.

31 Ibid., box 74, file 4, 1840s; Nicholas Marshall v John Meaney, 2 May 1843, HGNCC, writ no. 19.

32 Ibid., box 18, file 9, 1836; petition of Edward Shannahan to Judge E.B. Brenton, HGNCC, May 1836.

33 Ibid., box 21, file 5, 1848; petition of John Day to the HGNCC, 14 Nov. 1848.

34 Ibid., box 18, file 2, declaration of insolvency, 1 June 1837.

35 For examples of smaller insolvent operations see ibid., box 30, file 4, 1827; 'In the Insolvency of Wm. Mosdell,' 14 Nov. 1827, box 18, file 9,

1836; Thomas Godden v Thomas Sheehan, HGNCC, 18 April 1836, box 26, file 3, 1844; Thomas Ridley v John Parsons, Harbour Grace, 9 Nov. 1855, writ no. 28.

36 Ibid., box 21, file 6, 1848; insolvency of John Way, 23 Oct. 1848, writ no. 38.

37 Ibid., box 20, file 1, 1833; statement of the creditors of William Marshall, insolvent, in HGNCC, 19 Nov. 1833. For other examples, see box 59, HGNCC file 3, 1830–9; statement of debts and effects of James Shepherd in HGNCC, 14 Nov. 1832, box 20, file 1, 1833; estate of John Kennedy in HGNCC, 18 Nov. 1833, box 20, file 1, 1833; statement of the debts and effects of James Duggan in HGNCC, 19 Nov. 1833, box 55, file 2, 1830–9; petition of Howell and Cannon to Judge James Simms, HGNCC, 26 April 1847, box 21, file 5, 1848; petition of Moses Percey to the HGNCC, Brigus, 8 May 1848.

38 *Weekly Herald*, 11 March 1846.

39 Ibid., 11 April 1849.

CHAPTER 3 Household Agriculture

1 PANL, WMMS Corr., 1800–17, 971.8W1, no. 69; H. Busby to Rev. Robert Smith, Carbonear, 4 Jan. 1814; ibid., 1818–24, 971.8W2, no. 137, 'Observations &c on the Island of Newfoundland,' 1819.

2 R.G. Lounsbury, *British Fishery*, 3–25, 51–87, 145–92; Gillian T. Cell, *English Enterprise*, 7–21, 52–96; Anspach, *History of Newfoundland*, 334–412

3 Centre for Newfoundland Studies, Memorial University of Newfoundland, CO 194, microfilm box 677, vol. 37, 1783–7, fols. 125–33; 'Remarks of a Merchant in the Newfoundland fishery, 1781.'

4 Ibid., box 676, vol. 35, 1780–4, fols. 326–30; Archibald Buchanan to Keith Stuart, London, 5 April 1784.

5 Ibid., box 677, vol. 36, 1785–6, fols. 13–15; Gov. J. Campbell to Secretary of State Lord Sydney, St John's, 14 May 1785; vol. 37, 1783–7, fols. 138–40; petition to Lord Sydney from Poole merchants, 1785.

6 Ibid., box 679, vol. 40, 1798, fol. 31; Captain Crofton, Portsmouth, to Gov. W. Waldegrave, 10 Jan. 1798; fols. 137–8; Gov. W. Waldegrave to the Duke of Portland, *Agincourt* at sea, 30 Oct. 1798; box 680, vol. 43, 1801–3, fol. 3; C.W.M. Pole to the Duke of Portland, London, 12 Jan. 1801.

7 Ibid., box 680, vol. 43, 1801–3, fols. 189–99; T. Skerrett to Lord Pelham, St John's, 22 Sept. 1801.

8 Ibid., fols. 283–5; 'Committee of Council appointed for the considera-

tion of all Matters Relating to Trade and Foreign Plantations,' 2 March 1803; fol. 323; 'Extracts from a letter from Brigadier General Skerrett,' Newfoundland, 26 Oct. 1803; vol. 44, 1804–5, fols. 26–33; Gov. E. Gower to Earl Camden, Spithead, 19 Nov. 1804.

9 Ibid., vol. 44, 1804–5, fols. 50–3; Gov. E. Gower to Earl Camden, Hermitage, 24 Dec. 1804.

10 Ibid., box 681, vol. 45, 1806, fol. 253; Gov. E. Gower to Sir Stephen Cottrel, Hermitage, 9 June 1806.

11 Ibid., fol. 61, Gov. E. Gower; 'Observations on Certain Parts of His Majesty's Instructions to the Governor of Newfoundland,' 1806.

12 Ibid., box 682, vol. 49, 1809, fols. 52–64; Gov. T. Duckworth's Instructions and Observations, 1810; box 683, vol. 51, 1811, fols. 47–8; Duckworth to the Earl of Liverpool, London, 12 Nov. 1811; box 684, vol. 53, 1812, fols. 3–5; Duckworth to Lord Bathurst, *Antelope* at sea, 2 Nov. 1812.

13 PANL, P1/5, Duckworth Papers, microfilm reel M-3176, Rev. Edmund Violet, 'Memorandum on Agriculture in Newfoundland,' n.d., fols. 1153–7

14 CO 194, box 684, vol. 53, 1812, fols. 158–9; James Buller to H. Goulbourn, Whitehall, 15 Sept. 1812; fols. 170–4; Lewis A. Anspach to the CO, London, 9 Nov. 1812.

15 Ibid., box 684, vol. 54, 1813; Gov. R. Keats to Earl Bathurst, St John's, 23 June 1813; fols. 160–6; Keats to Bathurst, *Bellerophon* at Torbay, 18 Dec. 1813.

16 Prowse, *History of Newfoundland*, 404–5.

17 CO 194, box 686, vol. 59, 1817, fol. 82, Gov. F. Pickmore to Bathurst, London, 12 June 1817.

18 Ibid., fols. 92–3; John Brown, James Blackie, J. MacBraire, Peter W. Carter. and H. Brooking, justices of the peace, to Gov. F. Pickmore, St John's, 1 April 1817; fols. 110–12; Pickmore to Henry Goulbourn, Portsmouth, 20 July 1817; fol. 187; Pickmore to Bathurst, St John's, 22 Dec. 1817. Pickmore's correspondence unfortunately does not indicate how many paupers he might actually have sent out of Newfoundland.

19 Ibid., box 687, vol. 60, 1817, fols. 28–9; Thomas Lock to Henry Goulbourn, Office of the Committee of the Privy Council for Trade, Whitehall, 24 Sept. 1817.

20 Prowse describes the period, known as the '"The Winter of the Rals,"' as a time when 'gangs of half-famished, lawless men everywhere threatened the destruction of life and property' by store-breaking (see his *History of Newfoundland*, 405).

'Mob' was the term used by government officials, planters, and merchants to describe the gathering of fishing servants in their search for food that winter. George Rudé has well explored the use of the word as a pejorative by the preindustrial English and French ruling classes, who used it to describe the organization and protests of the crowd against food shortages, unfair food prices, or the nonobservance of customary rights. *Mob* thus became a way to dismiss such protests as those of the 'rabble,' not 'the people.' While not wishing to discount 'spontaneity in the origin, development, and climax of popular disturbance,' Rudé illustrated the unity of the essential beliefs, slogans, leaders, 'or some elementary or more developed form of organization' of pre-industrial disturbances. See Rudé, *Crowd in History*, 3–15, quotes from 244–5.

21 CO 194, box 687, vol. 60, 1817, fols. 185–6; petition of the Merchants of Bristol to Earl Bathurst, 13 March 1817; fols. 198–9, petition of the Merchants of Poole to Earl Bathurst, 12 Nov. 1819.

22 PANL, GN5/4/B/1, Minutes of the Harbour Grace Court of Session (hereafter cited as Sessions' Minutes), box 1, 1816–17 (Jan.–Sept.); 6 Nov. 1816, 2 Jan. 1817.

23 Ibid., William Pinsent, et al. to Surrogate Judge Thomas R. Toker, Harbour Grace, 18 June 1817.

24 Ibid., affidavit of Duncan McKellar, Port de Grave, 18 June 1817.

25 Ibid., testimony of Richard Shea and William Hampton, Port de Grave, 18 June 1817.

26 Ibid., testimony of George Best, 2 Jan. 1817.

27 Ibid., testimony of Charles Cozens, 16 June 1817.

28 CO 194, box 687, vol. 60, 1817, fols. 185–6; petition of the merchants of Bristol to Earl Bathurst, 13 March 1817. The merchants were Thomas Thorne & Co., Charles Nuttall & Co., William Danson, William Henderson, William Mullowney, and Bartholemew Henderson & Co.; fols. 211–12; Memorial of the Merchants of Poole, 7 March 1817. The merchants were Thomas Colbourne, George Garland, Christopher Spurrier & Co., Samuel and John Clark, John Slade & Co., Slade and Cox, Pack, Gosse and Fryer, George and James Kemp & Co., Sleat and Read, and Joseph Bird.

29 Ibid., box 687, vol. 60, 1817, fols. 189–91; James Henry Attwood to W.R.K. Douglas, London, 13 Aug. 1817.

30 See PANL, GN5/4/B/1, Sessions' Minutes, box 1, 1816–17, Jan.–Sept. 1817; Jn. Fergus v Peter Frey and John Griffin, Harbour Grace, 13 June 1817; Wm. Foley v Thomas Fogerty, Harbour Grace, 16 June 1817; *Rex* v

James Hedderson, Harbour Grace, 16 June 1817; *Rex* v William Hampton and William Lawless, Port de Grave, 18 June 1817.

31 PANL, GN5/1/B/1, Surrogate's Minutes, box 2, 1816-18; Record of the Surrogate, 29 May 1817.

32 Ibid., 1816-18; Decision of the Surrogate in the case of Keefe v Brennan, Carbonear, 24 April 1817.

33 CO 194, box 687, vol. 60, 1817, fols. 292-304; report from the Select Committee on the Newfoundland Trade, 19 June 1817, testimony of George Garland, J.H. Attwood, and George Kemp Jr.

34 Cadigan, 'Role of Fishing Ships.'

35 Carson, *Distress in Newfoundland,* 358-60.

36 Keith Matthews, 'The Class of '32: St John's Reformers on the Eve of Representative Government,' in Buckner and Frank, *Acadiensis Reader,* vol. 1, 212-26.

37 CO 194, box 688, vol. 64, 1821, fols. 121-3; Gov. C. Hamilton to Earl Bathurst, St John's, 4 Dec. 1821; fols. 129-30, petition of the inhabitants of St John's to Hamilton, 24 Oct. 1821; box 689, vol. 65, fol. 26; petition of the inhabitants of St John's to Hamilton, 6 May 1822; fols. 22-5, Hamilton to Bathurst, St John's, 6 May 1822; fols. 213-21, 'Report of the State of Newfoundland for the information of ... Earl Bathurst,' 1822.

38 Ibid., box 690, vol. 66, fol. 166, 1823, Hamilton to Bathurst, *Ranger* at sea, 28 Nov. 1823.

39 Ibid., box 691, vol. 68, 1824, fols. 78-80; 'Inquiry into the present state of the trade and fisheries of Newfoundland,' n.d., [1824?]; ibid., fols. 467-8; petition of the inhabitants of St John's to the House of Commons, 1824.

40 The Colonial Office limited the duration of the Judicature and Fisheries Acts to five years, but extended them to 1832 as it finalized plans for the granting of representative government to Newfoundland. See Alexander McEwen, 'Newfoundland Law of Real Property: The Origin and Development of Land Ownership,' PhD thesis, University of London, 1978, 106-8; Newfoundland Law Reform Commission, *Legislative History of the Judicature Act 1791-1988* (St John's 1989), 1-18.

41 McLintock, *Constitutional Government,* 145-92.

42 CO 194, box 692, vol. 70, 1825, fol. 186; Colonial Secretary E.B. Brenton to Archdeacon G. Coster, St John's, 11 Oct. 1825; fols. 188-91, 'Extract of a letter from Coster to Brenton,' Bonavista 15 Oct. 1825; fols. 192-6, Coster to Brenton, Bonavista, 18 Oct. 1825.

[margin annotations: "Bibliography not in bibliog", "not in biblios"]

CHAPTER 4 Women in Household Production

1 PANL, GN5/3/B/19, Harbour Grace Court Records (hereafter cited as HGCR), box 19, file 5, 'Information and Complaint of Catharine Gould,' Carbonear, HGNCC, 25 Sept. 1840.

2 Cohen, *Women's Work*, 6–12.

3 W. Gordon Handcock, 'English Migration to Newfoundland,' in Mannion, *Peopling of Newfoundland*, 15–48.

4 Handcock, 'Origins of English Settlement', 100–6.

5 Handcock, *there comes noe women*, 95.

6 This patriarchal family structure is examined by Ann Kussmaul, *Servants in Husbandry*, 3–25.

7 Upper Canadian inheritance law confirmed patriarchal authority over household property, including its transmission between generations. Male household heads rarely left property to women's control after their deaths, usually doing so only when no mature male heir existed. Even then, women usually were not allowed to alienate property from the deceased male's immediate descendants, particularly through remarriage. If sons were old enough, husbands commonly stipulated that wives and daughters be cared for by their sons without inheriting much property in their own right. As in Newfoundland, women could inherit some property for use in their lifetime, which reverted to male heirs upon their deaths. Daughters rarely inherited much property. Inheritance compelled women's final dependence on men's ownership of the means of production. See Cohen, *Women's Work*, 48–58.

As in Upper Canada, women in the northeastern United States lost the right to most property upon marriage based on English common law. Widows in Connecticut had a legal right to use one third of their deceased husbands' property as a life estate if their husbands' wills made no other provision. Women otherwise had few inheritance rights. Men would not allow women, through inheritance, to alienate any manner of household property from their patriarchal line. See Ditz, *Property and Kinship*, 119–39.

8 PANL, GN5/1/B/1, Surrogates' Minutes, box 1, 1789–92; ruling of Surrogate Judge Edward Packenham, 7 Oct. 1788.

9 Ibid.; Jane Mardon v LeCoux, 23 Oct. 1789.

10 PANL, GN5/2/A/1, Minutes of the Supreme Court of Newfoundland (hereafter cited as Supreme Court Minutes), box 2, 1811–18, Supreme Court Minutes 1817–18, fols. 35–6, 131–2; Nicholas Newell v Peter McPherson, 5 Aug. 1817, 15–16 Sept. 1817; box 3, 1818; Emma Gaden,

George Bayley, and Charles William Beverley v the administrator to the estate of Thomas Thistle, 1818.

11 See PANL, GN5/1/B/1, Surrogate's Minutes, Harbour Grace, box 1, Matthew Whelan v children of the late James Cole, 15 April 1795; Michael Boyce and the widow and children of Michael Mayne v Thomas Danson, 25 March 1797; petition of Jane Smith, 30 Nov. 1807; Hunt & Brien v Stapleton & Sons, 5 Nov. 1807; Jane Furneaux v Alfred Mayne, 29 April 1814; petition of Emmanuel Stone et al., 6 April 1815; box 2, William Pitts v John Billy, 20 Dec. 1817.

   PANL, GN5/3/B/19, HGCR, box 20, file 10, 1832; petition of James Black, Port de Grave, 7 May 1827; box 26, file 10, petition of James Halfpenny, Harbour Grace, 15 May 1827; petition of John Farrell, 29 May 1827; box 19, file 3, petition of John Roach on behalf of his mother, Ann Roach, 7 May 1831; box 27, file 12, petition of Catharine Gould, 2 May 1832; box 21, file 5; petition of Widow Donovan, 1848, file 12; petition of Catharine Donovan, 7 May 1852.

12 PANL GN5/3/B/19, HGCR, box 26, file 10; petition of Jane Smith, Bread and Cheese Cove, 4 June 1827.

13 Ibid., box 28, file 12, 1830–1; memorial of Mary Murphy to Judge E.B. Brenton, Harbour Grace, 8 Nov. 1838.

14 Ditz, *Property and Kinship*, 145–50.

15 PANL, GN5/1/B/1, Surrogate's Minutes, box 1, 1787–88; petition of Ann Brazill, 1 Oct. 1787; 1789–92, decision of Surrogate John Trigge, 23 Oct. 1789; Elizabeth Webber v Thomas Lewis, 15 Sept. 1792.

16 PANL, GN5/3/B/19, HGCR, box 65, file 1, 1820–9; petition of Eleanor Canty to Surrogate T. Nicholas, Harbour Grace, 8 Jan. 1822.

17 PANL, GN5/1/B/1, Surrogate's Minutes, box 1, 1789–92; Jane Cook for Ann Protch v Richard Brit, 23 Oct. 1790; 1793–7; Jane Cook v John Clements, 15 Nov. 1793; James MacBraire v Jane Cook, 4 Dec. 1793; 1787–1813; Mary Davis v Joseph Pynn, 25 Oct. 1793; Mary Davis v Simon Wells, 23 October 1794; Dinah White v John Thomey, 30 Oct. 1794; Amy Thistle v Jn. Churchill, 29 Feb. 1808; Martha Butt v George Butt, 9 Dec. 1813.

18 PANL, GN5/3/B/19, HGCR, box 20, file 10, 1832; petition of Ann Mugford to Chief Justice R.A. Tucker, Port de Grave, 21 May 1827; box 47, file 1, 1827–9; memorial of Ann Taylor to Tucker, Harbour Grace, 4 June 1829.

19 Cohen, *Women's Work*, 66–71.

20 PANL, WMMS Corr., 1818–24, 971.8W2, box 4-4, no. 137; 'Observations &c. on the island of Newfoundland,' March 1919; 1800–17, 971.8W1,

box 4-4, no. 64; Richard Taylor to Joseph Benson, Carbonear, 2 June 1812.

21 *Weekly Herald*, 27 Oct. 1852.

22 Head, *Eighteenth Century Newfoundland*, 4-5.

23 PANL, WMMS Corr., 1824-25, 971.8W5, box 4-4, no. 171; William Wilson to the Wesleyan Methodist Missionary Committee, Port de Grave, 6 Sept. 1825.

24 Tocque, *Wandering Thoughts*, 194-6.

25 PANL, GN5/1/B/1, Surrogate's Minutes, box 1, 1807-10; Thomas Farley v Kemps & Co., 14 Nov. 1808.

26 PANL, GN5/3/B/19, HGCR, box 20, file 2, 1833; petition of Patrick Loughlan to Judge E.B. Brenton, Harbour Grace, 31 Oct. 1833.

27 Ibid, box 20, file 1; statement of Joseph Pippy's voyage at Labrador, summer 1833, 10 Nov. 1833.

28 Ibid., box 50, file 3, 1821-47; petition of Mary Reed to Assistant Justice George Lilly, Harbour Grace, 1835.

29 PANL, GN5/1/B/1, Surrogate's Minutes, 1793-7; Richard Cornish v Grace and John Holmes, 6 Nov. 1794.

30 PANL, GN5/3/B/19, HGCR, box 59, file 3, 1830-9; deposition of Nancy Daw to the HGNCC, 14 May 1831.

31 'The Plea for Reform: The Case of James Landergan (1818),' in Neary and O'Flaherty, *By Great Waters*, 67-9. See also O'Flaherty, 'Lundrigan, James,' *Dictionary Canadian Biography*, vol. 6, 410.

32 Centre for Newfoundland Studies, Memorial University of Newfoundland, CO 194, box 688, vol. 64, 1821, fols. 6-8; 'Pleas in the Supreme Court of Saint John's Newfoundland, November 9th 1820.'

33 Cadigan, 'Whipping Them into Shape: State Refinement of Patriarchy among Conception Bay Fishing Families, 1785-1825,' in Neis and Porter, *Women's Lives*.

34 PANL, GN5/3/B/19, HGCR, box 21, file 1, 1828; complaint and examination of Edw. Janes, Harbour Grace, 17 Nov. 1827.

35 Ibid., box 47, file 11, 1832; information and complaint of James Sharp of Harbour Grace, chief constable and bailiff, before John Stark, JP, 8 Sept. 1836.

36 Ibid., box 21, file 1, 1828; oath of Mary Barry before Thomas Danson, JP, Harbour Grace, 2 Jan. 1828; oath of David Bansfield, 12 Jan. 1828.

37 Ibid., box 68, file 2, 1830-9; information and complaint of Ann Noel before Thomas Danson, JP, Carbonear, 19 Jan. 1835.

38 Ibid., box 38, file 7, 1836; 'Documents in the Case of William Hussey a dangerous Lunatic,' 29 March 1836.

39 Ibid., box 33, file 1, 1835–43; information and complaint of Charlotte Bradbury, wife of Charles Bradbury of Harbour Grace before Thomas Danson, JP, 28 Oct. 1844.

40 Ibid., box 41, file 14, 1851; information and complaint of Sarah Dalton, Western Bay, before Thomas Danson, JP, Harbour Grace, 7 Aug. 1851.

41 Ibid., box 47, file 11, 1832; information and complaint of Sarah Neary before John Stark, JP, Harbour Grace, 26 February 1838; box 59, file 5, 1850–9; minutes, Catharine Chitman v Edward Shanahan, assault & battery, Harbour Grace, 27 June 1853.

42 Ibid., box 47, file 10, 1832; examination of Mary Ryan, 9 Nov. 1833, box 33, file 2; 1835–43; papers in the case of Johanna Connors before Thomas Danson, JP, 19 Jan. 1835. Eliza Mills v John Burke, bastardy, HGNCC, 8 June 1847.

43 PANL, WMMS, Corr., 1818–24, 971.8W2, box 4-4, no. 66; John Walsh to Joseph Taylor, Conception Bay, 28 June 1819.

44 PANL, GN5/1/B/1, Surrogate's Minutes, 1787–8, Mary Cole v Stephen Hunt, 3 Oct. 1787, wage dispute. Hunt was a merchant who received all of the voyage of Ellison, but refused to allow Cole's wages.

45 PANL, GN5/3/B/19, HGCR, box 47, file 1, 1827–9; petition of Philip Meaney to Chief Justice R.A. Tucker, Harbour Grace, 5 June 1829; box 22, file 6; complaint of Mary Counsell, wife of James Counsell, Bay Roberts, 25 Aug. 1836; box 49, file 5, 1845; Ann French v Thomas French, 24 May 1845, writ no. 20.

46 Ibid., box 22, file 6; depositions of George Heater and Sophia Heater, Harbour Grace, 3 Nov. 1845.

47 Ibid., box 19, file 5; 'Information and Complaint of Bridget Cotter': Carbonear, HGNCC, 27 July 1844.

48 Ibid., box 54, file 2, 1830–9; examination of Susan Russell before Thomas Danson and James Power, JPs, Harbour Grace, 29 June 1839.

49 Ibid., box 61, file 5, 1850–9; minutes of Rebecca Slade v Ann Slade, HGNCC, 27 Aug. 1853.

50 Ibid., box 55, file 3, 1840–8; information and complaint of Edward Noftell before Thomas Danson and R.J. Pinsent, JPs, Harbour Grace, 2 Oct. 1845; box 66, file 1, 1820–9; information and complaint of Luke Micheton, Spaniards Bay, HGNCC, 14 Aug. 1834.

51 Light and Prentice, *Pioneer Gentlewomen*, 163–4. Notices reprinted in this collection are from Maritime newspapers.

52 Brenton, *Runaways*, 19–21.

53 *Gore Gazette*, 24 Jan. 1829.

54 CO 194, box 354, vol. 83, 1832, fols. 128–31; 'Extract from Mr. John McGoun's report of his proceedings whilst employed conveying Seed Potatoes and other relief to the Suffering Inhabitants of the Northern parts of Newfoundland.'

55 Cohen, *Women's Work*, 89–92.

56 *Carbonear Sentinel*, 26 Feb. 1839.

57 *Weekly Herald*, 3 March 1847.

58 Ibid., 28 June 1848.

59 Within another context, the manner in which women's subsistence activities could find little room in industrial capitalism is explored in Bradbury, 'Pigs, Cows, and Boarders,' 9–46.

60 On women in Newfoundland fishing households, see Porter, 'She Was Skipper,' 105–23, and Murray, *More than 50%*.

61 Ommer, 'Merchant Credit,' 188.

CHAPTER 5 The Legal Regime of the Fishery

1 Centre for Newfoundland Studies, Memorial University of Newfoundland, CO 194, box 679, vol. 41, 1771–98, fol. 71.

2 Cadigan, 'Merchant Capital,' *Journal Canadian Historical Association*, 17–42. On the law of master and servant, see Orren, *Belated Feudalism*, 84–114.

3 CO 194, box 676, vol. 34, 1776–7, fol. 33; unsigned petition from the 'principal merchants' of Newfoundland to the Secretary of State for the Colonies, 1777.

4 Ibid., box 678, vol. 38, 1788–91, fols. 290–302, quote from fol. 303.

5 Anspach, *History of Newfoundland*, 215–23.

6 CO 194, box 679, vol. 40, 1798, fol. 17–34; Captain Crofton, *Pluto*, to Gov. W. Waldegrave, Portsmouth, 10 Jan. 1798.

7 Ibid., vol. 41, 1771–98, fol. 64–70; 'A Brief State of the Evidence laid before the Committee of the House of Commons, upon the Newfoundland Trade and Fishery, in the last Session, 1793.'

8 Ryan, 'Fishery to Colony,' in Buckner and Frank, *Acadiensis Reader*, vol. 1, 147.

9 PANL, P1/5, Duckworth Papers, microplan, box no. R35.5, M-3176, fols. 3386–7; 'Opinion of G. Williams, T. Tremlett & Thos. Cooke on 15 G.III, c. 31,' 13 Nov. 1802.

10 CO 194, box 681, vol. 44, 1804–5, fol. 182; 'Extract of a Letter from the Rev. John Clinch Magistrate at Trinity to Joseph Frounsell Esq. Secretary to His Excellency Sir Erasmus Gower dated at Trinity 10 October 1805'; fol. 185; Gower to Viscount Castlereagh, St John's, 25 Oct. 1805.

11 Ibid., vol. 45, 1806, fols. 20–7; Gov. E. Gower, 'Explanatory Observations on the Accompanying Return of the Fishery and Inhabitants of Newfoundland ... 1804.'

12 Ibid., fols. 64–72; Gower, 'Observations on certain Parts of His Majesty's Instructions to the Governor of Newfoundland.'

13 Ibid., box 682, vol. 47, 1808, fols. 131–2; copy of the ruling on Cooke and Travers, insolvents, Conception Bay, Supreme Court, 1804.

14 Ibid., vol. 48, 1809, fols. 61–6; James McBraire et al. to Gov. J. Holloway, St John's, 19 Oct. 1809.

15 Ibid., fols. 81–4; charges of the Society of Merchants against Chief Justice Tremlett to Gov. J. Holloway, 1809.

16 Ibid., box 683, vol. 50, 1811, fols. 79–98.

17 Ibid., fols. 195–214; committee of the Society of Merchants to Gov. J. Duckworth, St John's, 24 Oct. 1811.

18 Ibid., box 684, vol. 53, 1812, fols. 93–6; minutes of the Committee of Trade and Foreign Plantations, Whitehall, 2 June 1812; box 686, vol. 59, 1817, fols. 10–20; Gov. F. Pickmore to Earl Bathurst, London, 7 Jan. 1817.

19 Ibid., box 687, vol. 60, 1817, fols. 249–56; J. Newart to Earl Bathurst, St John's, 14 March 1817.

20 Ibid., fols. 256–60.

21 Ibid., fols. 360–1.

22 Ibid., fols. 261–6.

23 Ibid., fols. 266–71.

24 Ibid., fols. 292–302.

25 PANL, GN5/2/A/1, Supreme Court Minutes, box 2, 1811–18; Supreme Court Minutes, July–Dec. 1817, Chief Justice Francis Forbes, fols. 183–6.

26 Ibid., Supreme Court Minutes, 1817; Philip Meany v. Thomas Pynn for £22.2.9. wages, 2 Dec. 1817.

27 The reasons are not clear concerning the brief decline of shares as a form of remuneration in the Newfoundland fishery. Gillian T. Cell notes that various shares arrangements existed in the early seventeenth century for the distribution of migratory ships' catches between owners and fishermen. Ralph G. Lounsbury's work on the later seventeenth and eighteenth centuries suggests that these share systems, preferred by West Country merchants, came into conflict with the desire of London and Bristol proprietary colonists to pay fixed wages in a resident fishery as a more 'capitalist' development of the fish trade. Keith Matthews emphatically disagreed with Lounsbury's view, pointing out that West Country

merchants neither opposed residence nor the use of wage labour, especially after 1660 when they relied on migratory bye-boat keepers' use of servants to actually catch fish. Matthews's argument seems to imply that wages overtook shares as part of the merchants' more general withdrawal from catching fish themselves. Harold Innis advanced the only explicit explanation, suggesting that fixed wages displaced the share system as the migratory fishery declined in the eighteenth century. Employers used fixed wages in an attempt to stem the migration of labourers from Newfoundland to New England.

See Cell, *English Enterprise*, 15–18; Lounsbury, *British Fishery*, 86–90; Keith Matthews, 'England-Newfoundland Fishery,' 120–70; Innis, *Cod Fisheries*, 151–5.

28 PANL, GN5/2/A/1, Supreme Court Minutes, box 2, Minutes, 1818; Stuarts and Rennie v David Walsh, 24 Jan. 1818.

29 Governors' annual reports to the Colonial Office indicated that shares increasingly dominated forms of remuneration given to servants after 1815. In Conception Bay, where planters used schooners in the declining north shore fishery and in the Labrador fishery, the proportion of servants hired on shares fluctuated between one-quarter and one-half of the total between 1815 and 1825. The high proportion of servants remaining on fixed wages probably reflects the more capital-intensive nature of the schooner fishery.

The governors reported that the planter fisheries of Trinity and Bonavista Bays were dominated by the use of sharemen, except those who did shore work. Merchants there, as well as at Fogo and Twillingate, still directly employed servants on fixed wages to catch and cure fish.

See CO 194, box 686, vol. 57, 1816, return of the fisheries and inhabitants, 10 Oct. 1814–11 Oct. 1815; box 696, vol. 9, 1817, 10 Oct. 1816–10 Oct. 1817; box 687, vol. 61, 1818, 10 Oct. 1817–10 Oct. 1818; box 688, vol. 62, 1819, fol. 149, 1818–19; box 688, vol. 64, 1821, 1820–1; box 690, vol. 5, 1822, fols. 120, 1821–2; box 690, vol. 66, 1823, fols. 172, 1823; box 692, vol. 70, 182, fols. 22, 33, 1824–5.

30 Ibid., box 688, vol. 63, 1820, fols. 134–141; Gov. C. Hamilton to Earl Bathurst, St John's, 27 Nov. 1820.

31 Ibid., fols. 250–6; Captain Nicholas to Lord Melville, Newfoundland, 18 Oct. 1820.

32 Ibid., vol. 64, 1821, fols. 65–8; Hamilton to Earl Bathurst, St John's, 26 July 1821.

33 Ibid., fol. 49, petition of the Inhabitants of St John's, 1821; fol. 129; peti-

tion of the inhabitants of St John's to Gov. Hamilton, 24 Oct. 1821;
Grand Jury to the Magistrates in general Session, St John's, 15 Oct. 1821;
Grand Jury to Chief Justice Francis Forbes, St John's, 13 Oct. 1821.

34 Ibid., box 689, vol. 65, 1822, fols. 234–5; James Dalton to James Butter-
worth, Liverpool, 4 April 1822.

35 Ibid., fols. 334–8; Robert Wilmot, secretary of state, remarks on the
'Laws Requisite' of Newfoundland, 1822.

36 Ibid., fols. 331–2; Hunt to Newman, Dartmouth, 6 April 1822; box 690,
vol. 66, 1823, fols. 340–3; James Dutton to R. Wilmot Horton, Liverpool,
24 July 1823.

37 Ibid., box 690, vol. 66, 1823, fols. 345–61; proposed bill 'For the Better
Administration of Justice in Newfoundland' with Gov. Hamilton's suggestions. vol. 67, 1824, fols. 49–51; Hamilton to Earl Bathurst, Spring
near Midhurst, 5 Feb. 1824.

38 Ibid., vol. 67, 1824, fols. 243–8; copy of the new Judicature Act, dated 17
June 1824.

39 Ibid., fol. 251; 'An Act to repeal several Laws relating to the Fisheries
... for Five Years.'

40 Ibid., box 691, vol. 68, 1824, fols. 112–36; J. Stephen Jr to Robert Wilmot
Horton, London, 19 March 1824.

41 Ibid., vol. 69, 1824, fols. 76–96; Wilmot Horton, 'Remarks upon the proposed Newfoundland Acts,' n.d.

42 Ibid., box 692, vol. 69, 1824, fols. 472–85; 'Inquiry into the present state
of the trade and fisheries of Newfoundland,' n.d.

43 Newfoundland Law Reform Commission, *Legislative History of the Judicature
Act 1791–1988* (St John's 1989), 1–18.

CHAPTER 6 Truck as Paternal Accommodation

1 PANL, GN5/3/B/19, HGCR, box 22, file 1; statement of William Lilly,
Harbour Grace, 12 Feb. 1848.

2 Ommer, 'Introduction,' in Ommer, *Merchant Credit*, 14.

3 Keith Matthews, 'England-Newfoundland Fishery,' 174–9; and Handcock,
there comes noe women, 233.

4 Ommer, 'Introduction,' 9–15.

5 Palmer, *Working-Class Experience*, 12–59.

6 PANL, GN5/3/B/19, HGCR, box 22, file 6, Peter Jordan v Joseph and
Robert Parsons, Harbour Grace, 22 Jan. 1832, writ no. 2, box 28, file 10,
1837; Richard McCarthy v John Keilley, Harbour Grace, 6 Nov. 1837,
writ no. 124, box 27, file 5, 1840; Thomas Snelgrove by his father Roger

v George Pelley, Harbour Grace, 7 Oct. 1840, writ no. 127, box 74, file 7, 1840; Thomas Bugden v Albert Pittman, Harbour Grace, 20 Dec. 1841, writ no. 112, box 74, file 1, 1840s; Patrick Houlahan v Richard Wedger, Harbour Grace, 17 Nov. 1842, box 20, file 6, 1853; John Dowle v James McCarthy, Harbour Grace, 15 Oct. 1853, writ no. illegible.

7 Ibid., box 22, file 7, miscellaneous; Charles Ives v Nathaniel Munden, Harbour Grace, 16 April 1832, writ no. 23, box 42, file 1, 1839–40; Thos. McNamara v Jonathan Taylor & Sons, Harbour Grace, 11 May 1840, writ no. 69, box 42, file 1, 1839–40; Francis King v Ridley, Harrison & Co., Harbour Grace, 7 May 1840, writ no. 2, box 47, file 2, 1835, 1846; Matthew Byrne v Punton & Munn, Harbour Grace, 2 May 1846, writ no. 10, box 21, file 5, 1848; John Jerrett v William Donnelly, Harbour Grace, 3 May 1848, writ no. 28, box 22, file 8; Michael Patten v Wm Bennett, Harbour Grace, 28 April 1832, writ no. 35.

8 Ibid., box 42, file 1, 1839–40; Patrick Power v Ridley, Harrison & Co., Harbour Grace, 7 May 1840, writ no. 65.

9 Ibid., box 20, file 10, 1832; Mary Rohan v Wm. Munden, Harbour Grace, 3 Nov. 1832, writ no. 183, box 28, file 5, 1837; Bridget Shannahan v Wm. Donnelly, Harbour Grace, 21 Nov. 1837, writ no. 163, box 28, file 5, 1837; Mary Power v Margaret McKie, executrix to the estate of Nicholas McKie of Carbonear, Harbour Grace, 20 Nov. 1837, writ no. 150, box 28, file 5, 1837; Hannah Piddle v Moses and Grace Gosse, Harbour Grace, 9 Nov. 1837, writ no. 133, box 74, file 1, 1840s; Alice Dunn v C. Watts, Harbour Grace, 4 Nov. 1842, writ no. 79.

10 Ibid., box 41, file 3, 1826; Peter Keefe v Robert Knox, Harbour Grace, 1 Dec. 1826, writ no. 241.

11 Ibid., box 30, file 5, 1827; Wm. Brennan v Thomas Pynn, Harbour Grace, 12 Nov. 1827, writ no. 84.

12 Ibid., box 41, file 3, 1826; Jeremiah Pumphrey v James Ball, Harbour Grace, 9 Dec. 1826, writ no. 276, box 18, file 5, 1833; John Mugford v Charles Boon, Harbour Grace, 1 Nov. 1833; writ no. 181, box 20, file 6, 1853; Francis Barrett v John Barrett, Harbour Grace, 10 Nov. 1853, writ no. 39, box 74, file 1, 1840s; Thomas and Patrick Healey v James Walsh, 23 Nov. 1842, writ no. 108.

13 Ibid., box 60, file 1, 1820–9.

14 Ibid., box 30, file 6, 1827.

15 Ibid., box 30, file 5, 1827; Thomas Melvin v Thomas Pynn, Harbour Grace, 10 Nov. 1827, writ no. 82, box 18, file 1, 1829; Patrick Rogers v Michael Joy and Richard Brennan of Carbonear, Harbour Grace, 1 Dec. 1829, writ no. 266, box 18, file 7, 1834; Wm. Walter v Wm. Brown of

Carbonear, 18 Nov. 1834, writ no. 345, box 18, file 2, 1829; Darby
Murphy v Laurence Maccassy of Cupids £3.2.0, the balance of his wages,
Harbour Grace, 16 Nov. 1829, writ no. 182; John Kehoe v Wm. Brennan
of Harbour Grace, £2.10.0, Harbour Grace, 20 Nov. 1829, writ no. 209,
file 6; Edmund Grinsel v Henry Davis, £8.5.4, the balance of his wages,
Harbour Grace, 14 Nov. 1833, writ no. 254, box 50, file 1, 1821–47; Thomas Couney v Edmund R. Danson, Harbour Grace, 21 May 1833, writ
no. 75, box 54, file 4, 1831; June Caval, administrator of the estate of the
deceased Tobin Caval, v David Power & Co. of Carbonear, the balance of
wages due, £4.1.0, Harbour Grace, 19 Nov. 1831, writ no. 241, box 41,
file 2, 1826; Joseph Verge v James R. Knight, Harbour Grace, 22 Dec.
1826, box 41, file 3, 1826; James Ryan v Thomas Pynn of Mosquito, for
£14 wages, Harbour Grace, 8 Dec. 1826, writ no. 273, box 56, file 3,
1830–9; Henry Bishop v James Legg and William Hefford, for £12
wages, 19 May 1835, writ no. illegible, box 18, file 8, 1834; Nicholas Dee
v John Shepherd, for £18 wages, Harbour Grace, 7 Nov. 1834, writ no.
257, box 50, file 1, 1821–47; Edmund Quinlan v John Leary and Thomas
Bulbert, for £10 wages, HGCR, 13 Nov. 1835, writ no. 194, box 50, file 1,
1821–47; Thomas Phelan v Robert Walsh, Harbour Grace, for £25 wages,
9 Nov. 1835, writ no. 173, box 27, file 4, 1840; Absolom Clarke v James
Howell, for £21 wages, Harbour Grace, 8 Nov. 1840, writ no. 147, box 27,
file 4, 1840; John Drake v William Winsor, Harbour Grace, for £15
wages 18 Nov. 1840, writ no. 176, box 74, file 4, 1840s; John Phillips v
David Whelan, Harbour Grace, 12 Dec. 1842, writ no. 114, box 38, file 9,
1844; Jacob Hall v Hurley and Scane, Harbour Grace, for £2 wages, 22
Oct. 1844, writ no. 49, box 38, file 9, 1844; John Hickey v Thomas
Deady, Harbour Grace, for £24 wages, 4 Nov. 1844, writ no. 60, box 49,
file 6, 1845; James Quinn v John Keilley, Harbour Grace, for £20 wages,
11 Nov. 1845, writ no. 53, box 56, file 5, 1840–9; Alfred Ash v Wm.
Yeatman, Harbour Grace, for £20 wages, 3 Dec. 1846, writ no. 62, box
26, file 4, 1855; Wm. Nicholas v Michael Donan, Harbour Grace, for
£34.12.4 wages, 5 Nov. 1855, writ no. 17.

16 PANL, GN5/1/B/1 Surrogates, Minutes, Harbour Grace, box 1,
1787–1818, book 1787–8; Daniel Hisney v his master Henry Widenham,
20 Oct. 1787. Hisney complained that Widenham would not give him a
copy of his account so that Hisney might understand why his master refused to pay the balance of his wages. See also M. Courney v William
Nile, 4 Oct. 1787 (servant M. Courney did not directly bring suit against
his master Peter Quinlan for his wages in 1787, but rather against William Nile, as agent for Quinlan's merchants, for refusing to pay Cour-

ney's account balance). Servant Thomas Hennessey v his master M. Kennedy, 4 Oct. 1787; and servant William Elliot v his master William Cochran, 10 Oct. 1787. Not all of the account-overcharge disputes involved fishing servants. Jeremiah Donny, agent to merchants Keefe & Sons, won his charge that they had overcharged him by £34.6.5. See Jeremiah Donny v Keefe & Sons, 13 Dec. 1787.

17 Ibid., box 1,1787–1818, book 1787–8; petition of Patrick Cochran to Edward Packenham, 19 Sept. 1787. Cochran sued 'his mistress' Eleanor Highland in 1787 because she short-paid him by a few shillings on his wages. The Surrogate Court ordered Highland to settle immediately with Cochran; see also the order of Surrogate Edward Packenham to District Constables, 23 Oct. 1787, and other cases: David Cushan v John Dowdle, 12 Nov. 1787. The other six cases are Joseph Keary v his master Dennis Conners, 26 Sept. 1787; Stephen Woodcock v his master Robert Morrisy, 1 Oct. 1787; Timothy Falway v his master William Pike, Clown's Cove, 4 Oct. 1787; Daniel Hisney v his master Henry Widenham, 20 Oct. 1787; M. Donnovan, Ed. Fitzgerald and others v William Keefe, 6 Nov. 1787; M. Quinlan, John Meagher and others v John Thomey, 6 Nov. 1787.

18 PANL, GN5/3/B/19, HGCR, box 41, file 3, 1826; Timothy Mulcahy v Robert Knox, Harbour Grace, 9 Dec. 1826, writ no. 275, box 18, file 2, 1829; John Connors v Wm. Brennan, Harbour Grace, 17 Nov. 1829, writ no. 194, box 18, file 5, 1833; Patrick Neale v Philip Walsh, HGCR, 8 Nov. 1833, writ no. 227, box 54, file 5, 1831; Michael Maratty v William Thistle, Harbour Grace, 22 Nov. 1831, writ no. 260.

19 Ibid., box 57, file 1, 1820–9; contract of Timothy Shea (or Sheehy) with Timothy Crimin, Brigus, 17 June 1826; box 58, file 1, 1820–29; deposition of Thomas Sheehy, Harbour Grace, 26 Oct. 1826.

20 Ibid., box 30, file 4, 1827; indenture of Thomas Pyne to Michael Barry, 31 May 1827, box 18, file 3, 1829; Patrick Moore v Slade, Elson & Co. and William Kehoe of Carbonear, Harbour Grace, 24 Nov. 1829, unsampled writ no. 228, box 28, file 10, 1837; Doyle v Shanahan and Walsh, Harbour Grace, 25 Nov. 1837, unsampled writ no. unknown, box 74, file 5, 1840s; complaint of Jonathan Percy before R.J. Pinsent, JP, HGNCC, 8 June 1841.

21 Ibid., box 38, file 10, 1844; William Fitzgibbon v Thomas Deady, Harbour Grace, 4 Nov. 1844, unsampled writ no. 61, box 59, file 5, 1850–9; William Donelly v Jeremiah Lee of Spaniard's Bay, Harbour Grace, 1 Nov. 1851, unsampled writ no. 31 with appended documents.

22 Ibid., box 20, file 1, 1833; Thomas Calvert v James Ciddihey and George Forward, box 18, file 3, 1829; John Landergan v Edmund Whiteway and

William Bennett, Harbour Grace, 17 Nov. 1829, writ no. 195, box 41, file 3, 1826; John Swift v Aaron Lewis and Alexander Campbell, Harbour Grace, 2 Dec. 1826, writ no. 247, box 18, file 6, 1833; James Brawders v Philip Walsh and Charles Nuttall, Harbour Grace, 16 Nov. 1833, writ no. 290. Servants occasionally sued only merchants: see box 19, file 2, 1829; Silvester Mahair v Thomas Chancey, Harbour Grace, 16 Nov. 1829, writ no. 184, box 19, file 6, miscellaneous; Richard Frayne v Charles Cozens, Harbour Grace, 11 Nov. 1829, writ no. 159, box 56, file 4, 1830–9; Michael Fitzgerald v Wm. Bennett, Harbour Grace, 17 Nov. 1830, writ no. 226, box 22, file 8, miscellaneous; Thos. Kenley v Nathaniel Munden, Harbour Grace, 6 Nov. 1842, writ no. 68.

23 Ibid., box 22, file 8, miscellaneous; James Brine v the assignees of the estate of H.W. Danson, Harbour Grace, 1 May 1832, writ no. 55 (for another example, see box 30, file 7, 1830); George Heater v John Fitzgerald, Harbour Grace, 4 Nov. 1830, writ no. illegible, box 20, file 9, 1832; James Conway v Abraham and Joseph Bartlett, Harbour Grace, 5 Nov. 1832, writ no. 173, box 20, file 9, 1832; John Hunt v Maurice Keene, Harbour Grace, 13 Oct. 1832, writ no. 162.

24 Ibid., box 18, file 6, 1833; John Healey v James Cuddihy, Harbour Grace, 16 Nov. 1833, writ no. 282, box 20, file 1, 1833; Charles Kavanagh v John Leary and William Bennett, Harbour Grace, 12 Nov. 1833, writ no. unknown.

25 Ibid., box 22, file 5, miscellaneous; Thomas Danson to Colonial Secretary James Crowdy, Harbour Grace, 22 Nov. 1833, box 48, file 3, 1831; oath of Wm. Bennett, Harbour Grace, 23 Nov. 1831, box 20, file 1, 1833.

26 Ibid., box 20, file 1, 1833.

27 Ibid., box 50, file 2, 1821-41; petition of Dennis Landergan & Michl. Miles, Harbour Grace, to Chief Justice R.A. Tucker, 8 Nov. 1832, box 18, file 4, 1833; Geoffrey Rielly v Benjamin Leary, Harbour Grace, 5 Nov. 1833, unsampled writ no. 200, box 20, file 1, 1833; 'A Statement of Wages due to Hamilton's Crew,' 1833.

28 Ibid., box 30, file 7, 1830; petition of James Coyne to Judge E.B. Brenton, HGNCC, 4 Nov. 1830.

29 Ibid., box 50, file 2, 1821-47; memorial of Garret Cavanagh to Judge A.W. DesBarres, 23 May 1831, box 50, file 2, 1821-47; petition of Patrick Flynn to Chief Justice R.A. Tucker, Harbour Grace, 9 Nov. 1832, box 55, file 2, 1830–9; memorial of Patrick McGrath to Judge A.W. DesBarres, Harbour Grace, 10 May 1831, with appended documents.

30 Ibid., box 68, file 3, 1830–9; petition of Owen Fitzgerald to Chief Justice R.A. Tucker, Harbour Grace, 9 Nov. 1832, box 61, file 4, 1850–9; infor-

mation and complaint of George Mills, Harbour Grace, fisherman, before R.J. Pinsent, JP, 6 Nov. 1854.

31 Ibid., box 68, file 2, 1830–9; examination of James Murphy, Carbonear, fisherman, before Thomas Danson, 15 Nov. 1833, box 69, file 3, 1830–9; information and complaint of Thomas Newell, agent of Slade, Elson & Co., Carbonear, 13 Nov. 1833, box 62, file 1, 1830–9; information and complaint of Thomas Ridley before Thomas Danson, JP, Harbour Grace, 14 Jan. 1837.

32 Ibid., box 49, file 10, 1855; information and complaint of Henry Thomey, planter, Harbour Grace, before R.J. Pinsent, JP, 30 Oct. 1855.

33 Ibid., box 47, file 10, 1832; memorial of William Bennett, merchant, Carbonear, to the magistrates of the Court of Session at Harbour Grace, 18 Feb. 1834.

34 Ibid., box 59, file 1, 1820–9; memorial of George Pynn, Musquito, to the magistrates of Harbour Grace, 10 Nov. 1828.

35 Ibid., box 49, file 12, miscellaneous; information and complaint of Charles Kennedy, planter, Harbour Grace, before Thomas Danson, JP, 19 June 1833, box 33, file 2, 1835-42; information and complaint of Martin Kelly, Harbour Grace, before Danson, 9 Sept. 1841, box 49, file 13, miscellaneous; order of Danson to the constables of the Northern District, 30 Oct. 1834.

36 Ibid., box 18, file 3, 1829; Pack, Gosse and Fryer v Thomas Hedderson, Harbour Grace, 21 Nov. 1829, unsampled writ no. 222.

37 Ibid., box 49, file 3, miscellaneous; Pack, Gosse and Fryer v Daniel Meaney, 1 Nov. 1840.

38 Morea's account is in ibid., box 38, file 10, 1844. Two of the writs issued by the Northern Circuit Court on behalf of Morea's servants turned up in the writ sample: box 38, file 11, 1844; Patrick Moratty v Timothy Morea, Harbour Grace, 4 Nov. 1844 (writ no. 55); and William Norea v Timothy Morea, Harbour Grace, 22 Oct. 1844 (writ no. 48). See also box 38, file 10, 1844; Richard Morea v Timothy Morea, Harbour Grace, 1 Nov. 1844 (writ no. 55) and Pack, Gosse and Fryer v Timothy Morea, Harbour Grace, 2 Nov. 1844 (writ no. 57).

39 Ibid., box 26, file 10, 1827; petition of Thomas Powell, Carbonear, to Chief Justice R.A. Tucker, 18 May 1827.

40 Ibid., box 30, file 4, 1827; petition of William Morey, planter, to Judge A.W. DesBarres, Carbonear, 15 Nov. 1827, box 18, file 2, 1829; John Hackett v John Mason, Harbour Grace, 30 Nov. 1829, unsampled writ no. 169.

41 Ibid., box 51, file 11, 1854; information and complaint of Jacob Nicholas,

planter, Harbour Grace, before R.J. Pinsent, JP, 24 Oct. 1854.

42 Ibid., box 18, file 18, 1833–6; petition of Michael Keefe, planter, Harbour Grace, to Judge E.B. Brenton, 19 June 1834, box 18, file 18, 1833–6; H.C. Watts to Judge Brenton, Harbour Grace, 20 June 1834.

43 Centre for Newfoundland Studies, Memorial University of Newfoundland, CO 194, box 697–8, vol. 79, 1829, fols. 199–200; W.A. Clarke to R.W. Hay, London, 9 March 1829.

44 *Weekly Herald*, 21 July 1847.

45 Ibid., 16 August 1848.

46 CO 194, box 686, vol. 57, 1816; returns of the fisheries and inhabitants, 10 Oct. 1814–11 Oct. 1815, box 686, vol. 59, 1817; 1816, box 687, vol. 61, 1817; 10 Oct. 1816–10 Oct. 1817. Box 687, vol. 61, 1818; 10 Oct. 1817–10 Oct. 1818. Box 688, vol. 62, 1819; 1818–19. Box 688, vol. 64, 1821; 10 Oct. 1819–20 Oct. 1820. Box 689, vol. 64, 1821, fol. 143; 1820–1. Box 689, vol. 65, 1822, fol. 120; 1821–2. Box 690, vol. 66, 1823, fol. 172; 1822–3. Box 692, vol. 70, 1825, fol. 22, 33; 1824, 1825.

47 PANL, GN5/3/B/19, HGCR, box 50, file 1, 1821–47; James L. Prendergast, agent for the trustees of the insolvent estate of Hugh William Danson, to Judge DesBarres, Harbour Grace, 20 May 1831, box 55, file 2, 1830–9; plea of Richard Taylor against Thomas Chancey and William W. Bulley, Harbour Grace, HGNCC, 1839, box 55, file 3, 1840–8; suit of George Pynn and Joseph Pike v William Udell, HGNCC, fall term, 1847.

48 Ibid., box 19, file 3, 1831 miscellaneous; John Gardiner to James Prendergast, Bristol, 12 March 1831, box 21, file 10, 1831–2.

49 Ibid., box 20, file 1, 1833; 'A Statement of Debts due to and of William Innott, 1833.'

50 *Weekly Herald*, 20 Oct. 1852.

51 Ibid., 2 March 1853.

52 Ibid., 7 Dec. 1853.

53 PANL, GN5/3/B/19, HGCR, box 22, file 1; testimony of Johnston Burrows, deputy sheriff of the Northern District, Harbour Grace, 14 Feb. 1848.

54 Ibid., testimony of James Power, Brigus, schoolmaster, 21 Feb. 1848; testimony of Joseph Cozens, Brigus, accountant, 21 Feb. 1848.

55 Ibid., testimony of Robert Leamon, 14 Feb. 1848; statement of John Saunders, fisherman, Brigus, 18 Feb. 1848; Patrick Morrissy, fisherman, Brigus, 19 Feb. 1848; John Lundregan, fisherman, Brigus, 19 Feb. 1848; Jeremiah Whelan, fisherman, Brigus, 19 Feb. 1848; Isaac Clarke, planter, 19 Feb. 1848; Caleb Whelan, planter, Brigus, 19 Feb. 1848; statement of

Thomas Stephens Jr, fisherman, Brigus, 18 Feb. 1848; statement of John Clarke Jr, fisherman, Brigus, 19 Feb. 1848

56 Ibid., statement of John Cole, fisherman, Colliers, 19 Feb. 1848.

57 Ibid., statement of Nathan Clarke, fisherman, Brigus, 19 Feb. 1848.

58 Ibid., statement of Thomas Stevens, fisherman, Brigus, 18 Feb. 1848.

59 Ibid., statement of Michael Merrigan, fisherman, Brigus, 17 Feb. 1848.

60 Ibid., statement of John Way Jr., fisherman, Brigus, 17 Feb. 1848.

61 Ibid., statement of John Sullivan, fisherman, Brigus, 17 Feb. 1848.

62 Ibid., statement of Eleanor Dunphy, fisherman, Brigus, 21 Feb. 1848; statement of John Dunphy, fisherman, Brigus, 21 Feb. 1848.

CHAPTER 7 Agriculture and Government Relief

1 PANL, GN2/A/1, colonial secretary's outgoing correspondence (hereafter cited as Col. Sec. Out. Corr.), vol. 46, 1846–7, fols. 382–3; Crowdy to Pinsent, St John's, 26 Oct. 1847; fols. 434–5; Crowdy to Pinsent, St John's, 17 Dec. 1847.

2 Centre for Newfoundland Studies, Memorial University of Newfoundland, CO 194, box 693, vol. 72, 1826, fols. 87–90; Gov. T. Cochrane to Bathurst, St John's, 30 Jan. 1826; box 695, vol. 74, 1827, fols. 29–32, 131–4; Cochrane to Bathurst, St John's, 13 Jan. 1827; Cochrane to Bathurst, St John's, 11 May 1827.

3 Ibid., box 695, vol. 75, 1827, fols. 175–209: Patrick Morris, *Remarks on Society.*

4 Ibid., box 696, vol. 76, 1828, fols. 60–2, 189–93, 197–9, 277–80; Pres. R.A. Tucker to William Huskinson, St John's, 25 Jan. 1828; Gov. T. Cochrane to Huskinson, 19 March 1828; Cochrane to Huskinson, London, 20 March 1828; Cochrane to Sir George Murray, St John's, 3 Sept. 1828.

5 *Public Ledger*, 28 July, 7 Aug., 9 Oct. 1829; 8 March 1831.

6 CO 194, box 533, vol. 81, 1831, fols. 34–40; Cochrane to Viscount Goderich, St John's, 3 Feb. 1831.

7 Ibid., fols. 53–63; reply of William Carson to queries of the Royal College of Physicians, 7 Feb. 1831.

8 Ibid., fol. 63.

9 Ibid., fols. 100–3.

10 Ibid., box 534, vol. 82, 1832, fols. 79–81; Pres. R.A. Tucker, 'Explanatory Statement to be Appended to the Blue Book for the year 1831.'

11 *Public Ledger*, 14, 21 March 1831.

12 Prowse, *History of Newfoundland*, 429.

13 There still exists no study of the manner in which Newfoundland at-
tained representative government. Inferences about British Colonial Office policy here are drawn from a number of sources: Manning, *British Colonial Government*, 474–542; Knorr, *British Colonial Theories*, 251–371; Shaw,
'introduction', *Great Britain and the Colonies*, 7; Burroughs, *Colonial Reformers*,
xxiv–xxv 24–5, 73–97; Burroughs, *British Attitudes*, 20–35, 120–49; Bayly,
Imperial Meridian, 193–247. Two works are helpful in understanding Colonial Office rule over the colonies: Knapland, *The British Colonial System*;
John W. Cell, *British Colonial Administration*.

The total disruption of the migratory fishery by the Napoleonic Wars
finally undermined opposition in the Colonial Office, and before it, the
Board of Trade, to colonial government and complete agricultural-land
ownership rights for Newfoundland. No longer could British officials
hope to preserve the migratory trade through opposition to a resident
fishery or to colonization. The Colonial Office had been debating what to
do about Newfoundland's relief problems since 1817, but it opposed colonial self-government for the island because it felt that Newfoundland's
economy could not shoulder the financial burden. The Newfoundland
Reformers' ally in the House of Commons, George Robinson, supported
by the radical Joseph Hume, gained government support for Newfoundland's self-government in 1831 in exchange for his vote for the Reform
Bill. See Leslie Harris, 'Representative Government,' 7–15; Ward, *Colonial
Self-Government*, 126–8. D.M. Young, *The Colonial Office*, 1–123; Buckner,
Responsible Government.

14 Smith, *An Inquiry*, 899–900.

15 CO 194, box 534, vol. 83, 1832, fols. 96–104; Pres. R.A. Tucker to Gode-
rich, St John's, 19 June 1832; fols. 124–7; Tucker to Goderich, St John's, 2
Aug. 1832; fols. 156–8; memorial of Joseph Mullowney to Pres. Tucker,
box 535, vol. 84, 1832, fol. 10; Vice Adm. E.G. Colpoys to George Elliot,
Bermuda, 21 June 1832; fols. 220–2; Benjamin Lester to Viscount Goderich, Poole, 24 April 1832; fols. 325–6; memorial of the merchants of
Poole to Goderich, Poole, 30 Jan. 1832; vol. 85, 1833, fols. 9–25; Gov.
Cochrane's first address to the first session of the House of Assembly,
1832; vol. 85, 1833, fols. 76–87, 207–9, 336–7; Cochrane to Goderich, St
John's, 18 Feb. 1833, Cochrane's report accompanying the Blue Books,
1832; Cochrane to Hay, St John's, 5 Nov. 1833, box 536, vol. 87, 1834,
fol. 84; address of the Legislative Council of Newfoundland to the King
in Council, 3 April 1834, box 537, vol. 88, 1834, fols. 11–12, 15–17; Cochrane to E.G. Stanley, St John's, 7 July 1834; Cochrane's report accompanying the Blue Books, 1833.

16 PANL, Cochrane Papers, Microfilm box 971.8C7, reel 7, fol. 2734, no. 13–14; E.G. Stanley to Cochrane, London, 28 May 1834.

17 PANL, GN2/2, incoming correspondence of the Colonial Secretary's office (hereafter cited as Col. Sec. In. Corr.), box 1832, vol. May-Aug., fols. 338–9; petition of Edmund Joseph Mullowney to R.A. Tucker, St John's, 23 July 1832; box 1833, vol. 2, fols. 112–21; H.J. Fitzgerald to Gov. Cochrane, 2 May 1833; box 1834–5, vol. Jan.-April 1835, fols. 205–8; petition of Thomas Danson to Gov. Prescott, Harbour Grace, 26 Feb. 1835. In this letter Danson summed up his recent efforts to get compensation.

18 *Public Ledger*, 15 April 1834.

19 *Sentinel*, 13 July 1837, 3 Aug. 1837, 7 June 1838, 11 Aug. 1838. The assembly and council fought over the former's attempts to pass road bills in their ongoing fight to control expenditure. The road bill controversy is fully explored in Gunn, *Political History*, 40–1.

20 Ibid., 12 March 1839.

21 Newfoundland, *Journal of the House of Assembly*, 2, vol. 4, 1839, 6, 15.

22 *Sentinel*, 2 July 1839, 17 March 1840, 24 March 1840.

23 Ibid., 21 April 1840, 14 March 1843, 8 Aug. 1843.

24 Crabb, 'Agriculture in Newfoundland,' 70–1; Patrick Morris, *Short Review*, 72–82.

25 *Public Ledger*, 17 Dec. 1841.

26 Newfoundland, *Journal of the House of Assembly*, 3, vol. 1, 1843, 51–2, 115, 368–70.

27 *Sentinel*, 30 July 1843.

28 *Weekly Herald*, 29 Jan. 1845.

29 Ibid., 26 Nov. 1845.

30 Ibid., 8 April 1846, 5 Aug. 1846, 27 Jan. 1846, 3 March 1847, 5 May 1847.

31 Ibid., 19 May June, 2 June 1847.

32 PANL, GN2/2, Col. Sec. Corr., 1847, fols. 29–31, C. Cozens to James Crowdy, Brigus, 18 January 1847.

33 Ibid., fols. 163–5, John Noad, president of the Newfoundland Agricultural Society, to James Crowdy, St John's, 25 March 1847.

34 *Weekly Herald*, 16 June 1847.

35 *Patriot*, 16 March 1842.

36 Ibid., 15 Sept. 1847; 6, 27 Nov. 1847.

37 PANL, GN2/2, Col. Sec. Corr., 1847, fols. 931–3, John Soaper to Gov. LeMarchant, St John's, 4 Nov. 1847.

38 *Weekly Herald*, 16 June 1847.

39 Ibid., 19 Jan. 1848.

40 Ibid., 19 April, 17 May 1848.
41 CO 194, box 561, vol. 128, 1847, fols. 108–13, 221–2, 226–7; petition of the merchants of Poole connected with the Newfoundland trade, to Earl Grey, Poole, 29 Nov. 1847; Gov. LeMarchant to Earl Grey, St John's, 1 May 1848.
42 PANL, GN5/3/B/19, HGCR, box 61, file 3, 1840–9; Colonial Secretary James Crowdy to the magistrates of Harbour Grace, St John's, 31 Jan. 1848.
43 CO 194, box 561, vol. 129, 1848, fols. 138–49, LeMarchant to Earl Grey, St John's, 4 May 1848.
44 Bayly, *Imperial Meridian*, 155–60.
45 Wynn, 'Spirit of Emulation,' *Acadiensis*, 5–51.
46 Buckner, 'Harvey, Sir John,' *Dictionary of Canadian Biography*, vol. 8, 374–84, and Waite, 'Sir John Gaspard,' *Dictionary of Canadian Biography*, vol. 10, 438–9.
47 PANL, GN2/A/1, Col. Sec. Out. Corr., vol. 44, 1842–3, fol. 78, fol. 212; vol. 45, 1843–6, fol. 289, fol. 323; vol. 46, 1846–7, fol. 85, Provincial Secretary James Crowdy to James Power, JP, Rev. H.G. Fitzgerald, J. Cummins, and W. Fawkner, 20 April 1842; Crowdy to Richard Barnes, 11 November 1842; Crowdy's 'Circular' to the magistrates, May 1845; Crowdy to Benjamin Sweetland, 4 July 1845; Crowdy to Rev. B. Smith, T. Waldron, and J. Smith of King's Cove, and David Candow, Thomas Mullowney, and John Sheers of Tickle Cove, 29 October 1846.
48 Ibid., vol. 46, 1846–7, fol. 172, Crowdy to Thomas Ridley and John Munn, 16 March 1847.
49 Ibid., fol. 177, fol. 188, fols. 242–4, Crowdy's 'Circular' to the magistrates of Newfoundland, 14 May 1847; see also Crowdy to Charles Cozens, JP, and the Revs. George Canter, D. Mackin, and J.S. Addy of Brigus, 25 March 1847; Crowdy to William Sweetland, JP, Rev. J.M. Wood, Mr. Scalan, and Jabez Ingham, 3 April 1847.
50 Ibid., fols. 367–8, Crowdy to Rev. James England, 13 Oct. 1857.
51 Ibid., fols. 378–9, Crowdy to Benjamin Sweetland, JP, and the magistrates of Bonavista, 21 Oct. 1847.
52 Ibid., fols. 382–3, fols. 397–8, 434–5, Crowdy to the Harbour Grace Commissioners of the Poor, 18 Dec. 1847, vol. 46, 1846–7, fols. 437–40; for LeMarchant's exchange with Pinsent, see Crowdy to Pinsent, 26 Oct., 9 Nov., 17 Dec. 1847.
53 *Weekly Herald*, 24 May 1848.
54 Ibid., 9 May 1849.

55 Ibid., 28 June 1848.

56 Ibid., 24 July, 25 Sept. 1850; 4 June 1851; 27 Oct. 1852; 13 July, 31 Aug., 7 Sept., 5 Oct. 1853; 22 Feb. 1854.

57 CO 194, box 661, vol. 134, 1851, fols. 51–9, LeMarchant to Earl Grey, St John's, 14 April 1851; box 662, vol. 136, 1852, fol. 157, LeMarchant to John J. Packington, St John's, 12 April 1852.

58 *Weekly Herald*, 24 May 1848.

59 Ibid., 24 Jan. 1849.

60 Ibid., 7 Feb. 1849.

61 Ibid., 4, 11 April 1849, 5 Sept. 1849; 9 Jan. 1850.

62 Ibid., 11 April 1849.

63 Ibid., 12 Oct., 9 Nov. 1853.

CHAPTER 8 Liberals and the Law

1 Gunn, *Political History*, 14–107.

2 McCann, 'Invention of Tradition,' *Journal of Canadian Studies*, 86–103; Buckner, 'Harvey, Sir John,' *Dictionary Canadian Biography*, vol. 8, 374–84; Hiller, 'Little, Philip Francis,' *Dictionary Canadian Biography*, vol. 12, 563–6; Leslie Harris, 'Parsons, Robert John,' *Dictionary Canadian Biography*, vol. 11, 673–4.

3 Gunn, *Political History*, 25–45.

4 PANL, GN2/2, Newfoundland, Col. Sec. Inc. Corr., box 1830, vol. 2, fols. 71–81; George Vandenhoff and Jn. Canning to Col. Sec. Ayre, Western Bay, 5 Oct. 1830; fols. 89–97; Charles Cozens to Ayre, Brigus, 7 Oct. 1830; fols. 43–5; Thomas Danson, Oliver St John, James Cawley, and William Stirling, JPs, to Ayre, Harbour Grace, 9 Oct. 1830; fols. 66–70; Robert J. Pinsent to Ayre, Port de Grave, 12 Oct. 1830; fols. 53–4; R. Otterhead to Gov. Cochrane, Heart's Content, 2 Oct. 1830; fols. 107–12; William Kelson to the magistrates at Trinity, 2 Oct. 1830; fols. 99–103; Samson Mifflin to Ayre, Bonavista, 11 Oct. 1830; fols. 142–7; George Frampton to Ayre, Greenspond, 20 Oct. 1830; fols. 241–8; Joseph Simms to Ayre, Twillingate, 23 Oct. 1830.

5 CO 194, box 534, vol. 82, 1831, fols. 24, 72–8; Supreme Court judges R.A. Tucker, A.W. DesBarres, and E.B. Brenton, 'the Report of our opinion upon the Judicature and Jurisprudence of this Colony,' St John's, 23 Aug. 1831.

6 Ibid., fols. 82–8.

7 Ibid., fols. 88–9. For those unfamiliar with the reference, 'kine' is the archaic plural of cow.

8 Ibid., box 535, vol. 84, 1832, fols. 261–98; James Simms, 'Report of HM Attorney General of Newfoundland on the Judicature Laws of that Colony, February 1832.'

9 Gunn, *Political History*, 17–19.

10 Romney, *Mr Attorney*, 63–82.

11 CO 194, box 539, vol. 93, 1835, fols. 24–6; Chief Justice H.J. Boulton to R.W. Hay, St John's, 7 Jan. 1835.

12 Gunn, *Political History*, 22–52.

13 The Newfoundland Liberals' overtures for fishing-servant support resembled the Wilkite liberals' earlier struggles in England against merchant-aristocratic domination of political power and the abuse of power by the British magistracy. In their appeal for popular support, the Wilkites made much of the lower order's rights as British subjects to have the rule of English common law fairly applied to them. True Englishmen did not have to subject themselves to the arbitrary exercise of authority on the part of the judiciary.

   See Brewer, *Popular Politics*, 163–91; Rudé, *Wilkes and Liberty*, 176–83; Brewer, 'The Wilkites and the Law 1763–74: A Study of Radical Notions of Governance,' in Brewer and Styles, *Ungovernable People*, 132–71.

14 CO 194, box 538, vol. 90, 1835, fols. 319–36; memorial of the inhabitants of St John's.

15 Ibid., box 544, vol. 98, 1837, fols. 213–14; petition of the House of Assembly to Queen Victoria, 18 Oct. 1837, William Carson – Speaker.

16 On the nature of Irish immigration to Upper Canada, see Akenson, *Irish in Ontario*, 3–47; for Irish support of the Tories, for their own purposes, see 139–201. On Roman Catholic hierarchical support of the Tories, see Kerr, 'Orange and Green' *Ontario History*, 34–42.

17 Senior, 'Boulton, Henry John,' *Dictionary Canadian Biography*, vol. 9, 70.

18 On the political roles of the Roman Catholic bishops in Newfoundland, see Howley, *Ecclesiastical History*, 170–85, 255–330; Byrne, *Gentlemen-Bishops*, 4–27; Lahey, *James Louis O'Donel*; Lahey, 'Ewer (Yore), Thomas Anthony'; 'Power, John'; and 'Scallan, Thomas'; *Dictionary Canadian Biography*, vol. 6, 243–4, 613–14, 690–4.

19 My interpretation of the O'Connellite courting of fishing servants by Fleming and the Liberals comes from Lahey, 'Fleming, Michael Anthony,' *Dictionary Canadian Biography*, vol. 7, 292–99; Philip McCann, 'Bishop Fleming and the Politicization of the Irish Roman Catholics in Newfoundland, 1830–1850,' in Murphy and Byrne, *Religion and Identity*, 81–98; Keith Matthews, 'The Class of '32: The Newfoundland Reformers on the Eve of Representative Government,' in Buckner and Frank, *Acadiensis Reader*,

vol. 1, 212–26. My perspective is informed by Wilson, 'Irish in North America,' 127–32.

20 Little, 'Collective Action,' 7–35.

21 Howley, *Ecclesiastical History*, 329–30.

22 CO 194, box 544, vol. 99, 1837, fols. 146–52; Chief Justice H.J. Boulton to Gov. Prescott, St John's, 27 Sept. 1837.

23 Ibid., fols. 153–66; Boulton to Prescott.

24 Ibid., fols. 171–85.

25 *Patriot*, St John's, 30 June 1835.

26 Ibid., 13 October, 3 November 1835.

27 Ibid., 3 Nov. 1835.

28 Ibid.

29 Ibid., 24 Nov. 1835.

30 Ibid., 1 Dec. 1835.

31 *Public Ledger*, 5 Feb. 1836.

32 Ibid., 5 Feb. 1836.

33 CO 194, box 541, vol. 95, 1836, fols. 50–5; petition of the inhabitants of Newfoundland to the King in Council, 1836; fols. 62–4; J.V. Nugent to Joseph Templeman, acting Col. Sec., St John's, 12 July 1836.

34 *Patriot*, 19 Jan. 1835.

35 Ibid., 9 Feb. 1836.

36 Ibid., 16 Feb. 1836.

37 Ibid., 21 May 1836.

38 Ibid., 3 Dec. 1836.

39 Ibid., 11 Feb. 1837. Parson's commitment to his version of responsible government appears to have stemmed from, or at least to have been reinforced by, his examination of Upper Canadian affairs in the *Patriot*. While this may partially be a result of Boulton's origins, Newfoundland interest in Upper Canada rather than the Maritimes had much deeper roots. Certain officials, like Governor Harvey, may have been assigned to the island by the Colonial Office, or John Kent might have spoken with Joseph Howe about responsible government (misunderstanding the latter in the process; see Wells, 'Struggle for Responsible Government,' 15–17), but there is little reason to believe that this was because of any special interregional links in the politics or society of what has become known as Atlantic Canada. The previous chapter shows that planters who left Newfoundland searched out better lives in the American midwest, still very much the settlement frontier of North America.

The idea of responsible government as party government began to gain public prominence throughout British North America at this time, becoming visible with Upper Canada's very public battle between Bond Head and the radical Reformers during the 1836 general election, when Boulton was under heavy attack from the Liberals. See Buckner, *Responsible Government*, for a description of the general phenomenon. The events in Upper Canada are examined in Cadigan, 'Paternalism and Politics,' 319–47.

40 PANL, GN2/2, Col. Sec. Corr., box 1837, vol. 1 (Jan.-April) 1837, fols. 96–105; Chief Justice H.J. Boulton to Gov. Prescott, St John's, 25 Jan. 1837.

41 Ibid., vol. 2, fols. 45–50; Boulton to Prescott, St John's, 28 July 1837.

42 CO 194, box 545, vol. 103, 1838; 'Complaints against Chief Justice Boulton'; 'Report of Committee of the whole House of Assembly on the present state of Justice in Newfoundland'; J.V. Nugent, Chair, 10 October 1837; 'Report of Committee of the whole,' Nugent, 10 Oct. 1837, fol. 4; ibid., vol. 102, 1838, fol. 407; 'Copy of Addresses received at the Colonial Office from the ... Assembly of Newfoundland,' William Carson, Speaker, 25 Oct. 1838.

43 PANL, CO 194, reel A50 (A-3-1), vol. 103, 1838, fol. 263; petition of Patrick Morris to Lord Glenelg, St John's, 26 April 1838.

44 Ibid., fol. 266; Morris Petition, 1838.

45 PANL, GN5/3/B/19, HGCR, box 37, file 8, 1845: James Simms to R.J. Pinsent, JP, St John's, 17 Jan. 1845; box 76, file 5, 'An Act regulating the service of Merchant seamen engaged in Nfld 1837.'

46 O'Flaherty, 'Lundrigan, James,' *Dictionary Canadian Biography*, vol. 6, 410.

47 PANL, CO 194, A50 (A-3-1), vol. 102, fol. 266; Morris petition, 1838. Morris' statement unwittingly showed that the wage and lien system was a custom of the migratory fishery.

48 Newfoundland, *Journal of the House of Assembly*, 2, vols. 2 & 3, 1838, appendix 2, 'Memorial of Patrick Morris to the Right Honourable Lord Glenelg,' 26 April 1838, 98–104. This is an unedited version of Morris's memorial.

49 PANL, GN5/2/A/1, Supreme Court of Newfoundland, box 2, minutes 1817–18, July-Dec. 1817, Chief Justice Francis Forbes, fols. 183–9, appeal of trustees of Crawford & Co. v Cunningham & Bell, box 2, 1811–15/ 1817–18, Minutes, July-Dec. 1817, Chief Justice Francis Forbes, fols. 261–3; Keefe v trustees of Shannon & Co., 8 Dec. 1817; ibid., box 5, Minutes, 1823–24, Chief Justice R.A. Tucker; Brehaut & Sheppard v Trustees of LesMessuriers estate, for debt £2887.2.4, 10 Nov. 1823.

50 CO 194, box 546, vol. 103, 1838, fols. 325–46; 'In the Privy Council. In the Matter of the Complaint of the House of Assembly of Newfoundland

against the Honourable H.J. Boulton, Chief Justice of Newfoundland. Case of the Chief Justice.'

51 Gunn, *Political History*, 45.
52 *Public Ledger*, 3 March 1840; 26 Jan. 1841.
53 *Patriot*, 21, 28 Nov. 1840, 19 Dec. 1840.
54 *Public Ledger*, 12 March 1841.
55 Newfoundland, *Journal of the House of Assembly*, 2, vol. 6, 1841, Appendix, pp. 82–3; James Simms to Col. Sec. Crowdy, St John's, 7 March 1841; H.A. Emerson to Crowdy, St John's, 3 March 1841.
56 *Public Ledger*, 2 April 1841.
57 Liberal leadership had fractured in 1840 over a by-election contest in St John's. James Douglas, a Liberal Presbyterian merchant, first declared his candidacy, supported by Carson and Parsons, only to be challenged by Irish Catholic merchant Laurence O'Brien, supported by John Kent, J.V. Nugent, and Bishop Fleming. O'Brien won the campaign, but alienated Protestant Liberals like Parsons. Shortly after, Douglas began to assert that his followers were true Newfoundland natives, not Irish-born foreigners. The Newfoundland Natives Society emerged from public sentiment similar to Douglas's, supported by Parsons. The *Patriot* editor hoped to build an alliance of merchants and anti-clerical liberals to gain constitutional change. The Irish Catholic faction of the Liberals, which dominated the Assembly, cut off Parson's patronage as the Assembly printer. This punishment, and Natives Society ambiguity on the issue of constitutional reform, led Parsons back to the Liberal fold. See Budden, 'Newfoundland Natives Society,' 11–52.
58 *Patriot*, 13 Feb. 1841, 10 March 1841.
59 *Public Ledger*, 3 March 1843.
60 *Patriot*, 18 Jan. 1843.
61 Ibid., 20 March 1844.
62 Philip Buckner, 'Harvey, Sir John,' 375–84.
63 *Patriot*, 6 Nov. 1844.
64 Newfoundland, *Journal of the House of Assembly*, 3, vol. 3, Appendix, Fisheries, 202; Thomas Norton, George Lilly, and James Simms to the Governor, St John's, 26 March 1845. See also *Weekly Herald*, 23 April 1845.
65 *Patriot*, 30 April 1845.
66 Ibid., 20 Nov. 1844.
67 *Patriot*, St John's, 15 Jan. 1845. The rally's proceedings are summarized in the *Weekly Herald*, 8 Jan. 1845; 29 Jan. 1845. Too few details of these rallies exist to allow a full analysis of their composition and effect.
68 *Patriot*, 4 Feb. 1846.

69 Gunn, *Political History*, 128–40.

70 Robert M. Lewis, 'The Survival of the Planters' Fishery in Nineteenth and Twentieth Century Newfoundland,' in Ommer, *Merchant Credit*, 110–11.

CONCLUSION

1 Peter Neary, 'The French and American Shore Questions as Factors in Newfoundland History,' in Hiller and Neary, *Newfoundland*, 95–122.

2 Alexander, 'Development and Dependence in Newfoundland, 1880–1970,' in Sager, Fischer, and Pierson, *Atlantic Canada and Confederation*, 3–31.

3 Hatton and Harvey, *Newfoundland*, ix–x, 115–16.

4 Talbot, *Newfoundland*, 12–13.

5 James Murray, *The Commercial Crisis in Newfoundland: Cause, Consequence and Cure* (St John's 1895), 3.

6 McDonald, 'To Each His Own,' passim.

7 Smallwood, *New Newfoundland*, 231–3.

8 Smallwood, *Dr. William Carson*, 9.

9 Alexander, *The Decay of Trade*, 1–18, 128–57; J.K. Hiller, 'Newfoundland Confronts Canada, 1867–1949,' in E.R. Forbes and D.A. Muise, eds., *The Atlantic Provinces in Confederation* (Toronto: University of Toronto Press; Fredericton: Acadiensis Press, 1993), 349–81; David Ralph Matthews, *Controlling Common Property*, 38–65.

Bibliography

(absence of abbreviations list in awkward)

PRIMARY SOURCES

I. Manuscripts

A. Centre for Newfoundland Studies Great Britain. *(not some catalogue or in footnotes p. 185)*

Colonial Office Records, 1775–1855. CO 194, microfilms B-533–561, B-659–668, B-674–698.

B. Provincial Archives of Newfoundland and Labrador.

Newfoundland. Court of Session, Northern Circuit, 1826–55, GN5/4/B/1. *(which is HGCR + which HGNCC? p. 187)*
Newfoundland. Harbour Grace Court Records, 1817–55, GN5/3/B/19.
Newfoundland. Incoming Correspondence of the Colonial Secretary's Office, 1826–55, GN2/2.
Newfoundland. Outgoing Correspondence of the Colonial Secretary's Office, 1826–55, GN2/1/A.
Newfoundland. Supreme Court, Minutes, 1811–55, GN5/2/A/1.
Newfoundland. Supreme Court, Northern Circuit, Minutes, 1826–55, GN5/2/B/1.
Newfoundland. Surrogate Court, Minutes, 1785–1826, GN5/1/B/1.
Sir Thomas Cochrane Papers, P1/7.
Sir Thomas Duckworth Papers, P1/5.
Society for the Propagation of the Gospel in Foreign Parts. 'C' Series. MSS Canada, Nova Scotia (Newfoundland), 1787–1855.
Wesleyan Methodist Missionary Society, Newfoundland Correspondence, 1815–55.

II. Newspapers

The *Gore Gazette*, Ancaster, Upper Canada, 1829.
The *Patriot*, St John's, 1833–55.
The *Public Ledger*, St John's, 1827–55.
The *Royal Gazette*, St John's, 1810–55.
The *Sentinel*, Carbonear, 1836–45.
The *Weekly Herald*, Harbour Grace, 1845–54.

III. Published Documents

Newfoundland. *Abstract Census and Return of the Population, 1845.*
Newfoundland. *Abstract Census and Return of the Population, 1857.*
Newfoundland. *Journals of the House of Assembly, 1832–55.*
Newfoundland. *Population Returns, 1836.*

IV. Books

Anspach, Lewis A. *A History of the Island of Newfoundland: Containing a Description of the Island, the Banks, the Fisheries, and Trade of Newfoundland and the Coast of Labrador.* London: T&J Allen, 1819.
Bonnycastle, Sir Richard. *Newfoundland in 1842.* London: Henry Colburn, 1842.
Carson, William. *Reasons for Colonizing the island of Newfoundland.* Greenock, Scotland: William Scott, 1813.
– *Distress in Newfoundland.* St John's, Nfld.: 1817.
Morris, E.P., ed. *Decisions of the Supreme Court of Newfoundland: The Reports, 1817–1828.* St John's, Nfld.: King's Printer, 1901.
Morris, Patrick. *Remarks on the State of Society, Religion, Morals, and Education at Newfoundland.* London: A. Hancock, 1827.
– *A Short Review of the History, Government, Constitution, Fishery and Agriculture of Newfoundland.* St John's, Nfld.: J. Woods, 1848.
Reeves, John. *History of the Government of the Island of Newfoundland* (1793). East Ardsley: S.R. Publishers, 1967.
Tocque, P. *Wandering Thoughts or Solitary Hours.* London: Thomas Richardson & Son, 1844.

SECONDARY SOURCES

I. Books

Akenson, D.H., ed. *Canadian Papers in Rural History.* Vol. 2. Gananoque: Langdale Press, 1980.
– *Canadian Papers in Rural History.* Vol. 5. Gananoque: Langdale Press, 1986.
– *The Irish in Ontario: A Study in Rural History.* Montreal and Kingston: McGill-Queen's University Press, 1984.
Alexander, David G.. *The Decay of Trade: An Economic History of the Newfoundland Saltfish Trade, 1935-1965.* St John's: Institute of Social and Economic Research, Memorial University of Newfoundland, 1977.
– *Atlantic Canada and Confederation: Essays in Canadian Political Economy.* Edited by Eric W. Sager, Lewis R. Fischer, and Stuart O. Pierson. Toronto: University of Toronto Press, 1983.
Aston, T.H., and C.H.E. Philpin, eds. *The Brenner Debate: Agrarian Class Structure and Economic Development in Pre-Industrial Europe.* Cambridge: Cambridge University Press, 1976, 1978, 1979, 1982, 1985.
Baldwin, Douglas. *Land of the Red Soil: A Popular History of Prince Edward Island.* Charlottetown: Ragweed Press, 1990.
Barron, Hal, S. *Those Who Stayed Behind: Rural Society in Nineteenth-Century New England.* Cambridge: Cambridge University Press, 1984.
Bayly, C.A. *Imperial Meridian: The British Empire and the World, 1780-1830.* London: Longman, 1989.
Beck, J. Murray. *Joseph Howe,* Vol. 1. *Conservative Reformer, 1804-1848.* Montreal and Kingston: McGill-Queen's University Press, 1982.
Berg, Maxine, Pat Hudson, and Michael Sonenecher, eds. *Manufacture in Town and Country before the Factory.* Cambridge: Cambridge University Press, 1983.
Bois, Guy. *The Crisis of Feudalism: Economy and Society in Eastern Normandy c. 1300-1550.* Cambridge: Cambridge University Press, 1984.
Bottomore, Tom, Laurence Harris, V.G. Kiernan, and Ralph Milliband, eds. *A Dictionary of Marxist Thought.* Cambridge, Mass.: Harvard University Press, 1983.
Brenton, Myron. *The Runaways.* Boston: Little, Brown, 1978.
Brewer, John. *Party, Ideology and Popular Politics at the Ascension of George III.* Cambridge: Cambridge University Press, 1976.
Brewer, John, and John Styles, eds. *An Ungovernable People.* New Brunswick, NJ: Rutgers University Press, 1975.
Brookfield, H.C. *Colonialism, Development and Independence.* Cambridge: Cambridge University Press, 1972.

– *Interdependent Development*. Pittsburgh: University of Pittsburgh Press, 1975.

Buckner, Phillip A. *The Transition to Responsible Government: British Policy in British North America, 1815–1850*. Westport, Conn.: Greenwood Press, 1985.

Buckner, Phillip A., and David Frank, eds. *The Acadiensis Reader*. Vol. 1, *Atlantic Canada before Confederation*. Fredericton, NB: Acadiensis Press, 1985.

– *The Acadiensis Reader*. Vol. 2, *Atlantic Canada after Confederation*. Fredericton, NB: Acadiensis Press, 1989.

Buckner, Phillip A., and John G. Reid, eds. *The Atlantic Region to Confederation: A History*. Toronto: University of Toronto Press; Fredericton, NB: Acadiensis Press, 1994.

Burroughs, Peter, ed. *The Colonial Reformers and Canada: 1830–1849*. Toronto: McClelland and Stewart, 1969.

– *British Attitudes towards Canada, 1822–1849*. Scarborough, Ont.: Prentice-Hall, 1971.

Byrne, Cyril, ed. *Gentlemen-Bishops and Faction Fighters*. St John's, Nfld.: Jesperson Press, 1984.

Cell, Gillian T. *English Enterprise in Newfoundland 1577–1660*. Toronto: University of Toronto Press, 1969.

Cell, John W. *British Colonial Administration in the Mid-Nineteenth Century: The Policy-Making Process*. New Haven: Yale University Press, 1970.

Clement, Wallace, and Glenn Williams, eds. *The New Canadian Political Economy*. Montreal and Kingston: McGill-Queen's University Press, 1989.

Cohen, Marjorie Griffin. *Women's Work, Markets, and Economic Development in Nineteenth-Century Ontario*. Toronto: University of Toronto Press, 1988.

Corrigan, Philip, and Derek Sayer. *The Great Arch*. New York: Basil Blackwood, 1985.

Craig, Gerald M. *Upper Canada: The Formative Years*. Toronto: McClelland and Stewart, 1963.

Denoon, Donald. *Settler Capitalism: The Dynamics of Dependent Development*. Oxford: Clarendon Press, 1983.

Ditz, Toby L. *Property and Kinship: Inheritance in Early Connecticut, 1750–1820*. Princeton: Princeton University Press, 1986.

Dobb, Maurice. *Studies in the Development of Capitalism*, 5th ed. New York: International Publishers, 1957.

Duffy, Ian P.H. *Bankruptcy and Insolvency during the Industrial Revolution*. New York: Garland Publishers, 1985.

Easterbrook, W.T., and M.H. Watkins, eds. *Approaches to Canadian Economic History*. Toronto: McClelland and Stewart, 1967.

Ellison, Suzanne. *Historical Directory of Newfoundland and Labrador Newspapers, 1807–1987*. St John's: Memorial University of Newfoundland Library, 1988.

Fischer, Lewis, and Eric W. Sager, eds. *The Enterprising Canadians: Entrepreneurs and Economic Development in Eastern Canada, 1820–1914.* St John's: Maritime History Group, Memorial University of Newfoundland, 1979.

Flaherty, David H., ed. *Essays in the History of Canadian Law.* Vol. 1. Toronto: University of Toronto Press, 1981.

Forbath, William E. *Law and the Shaping of the American Labor Movement.* Cambridge: Harvard University Press, 1991.

Forbes, Ernest. *Challenging the Regional Stereotype: Essays on the 20th Century Maritimes.* Fredericton, NB: Acadiensis Press, 1989.

Fox-Genovese, Elizabeth, and Eugene D. Genovese. *Fruits of Merchant Capital: Slavery and Bourgeois Property in the Rise and Expansion of Capitalism.* New York: Oxford University Press, 1984.

Frank, André Gunder. *World Accumulation, 1492–1789.* New York: Monthly Review Press, 1978.

– *Dependent Accumulation and Underdevelopment.* New York: Monthly Review Press, 1979.

Furtado, Celso. *Economic Development of Latin America.* Cambridge: Cambridge University Press, 1970, 1976.

Gagan, David. *Hopeful Travellers: Families, Land, and Social Change in Mid-Victorian Peel County, Canada West.* Toronto: University of Toronto Press, 1981.

Gentilcore, R. Louis, ed. *Historical Atlas of Canada.* Vol. 2, *The Land Transformed, 1800–1891.* Toronto: University of Toronto Press, 1993.

Gilmour, James M. *Spatial Evolution of Manufacturing in Southern Ontario, 1851–1891.* Toronto: University of Toronto Press, 1972.

Goody, Esther N., ed. *From Craft to Industry: The Ethnography of Proto-industrial Cloth Production.* Cambridge: Cambridge University Press, 1982.

Gosse, Edmund. *The Life of Philip Henry Gosse.* London: Kegan, Paul, 1890.

Greer, Allan. *Peasant, Lord and Merchant: Rural Society in Three Quebec Parishes, 1740–1840.* Toronto: University of Toronto Press, 1985.

– *The Patriots and the People: The Rebellion of 1837 in Rural Lower Canada.* Toronto: University of Toronto Press, 1993.

Gunn, Gertrude E. *The Political History of Newfoundland, 1832–1864.* Toronto: University of Toronto Press, 1966.

Hahn, Steven. *The Roots of Southern Populism.* New York: Oxford University Press, 1983.

Hahn, Steven, and Jonathan Prude, eds. *The Countryside in the Age of Capitalist Transformation.* Chapel Hill, NC: University of North Carolina Press, 1985.

Handcock, W. Gordon. *Soe longe as there comes noe women: Origins of English Settlement in Newfoundland.* St John's: Breakwater, 1989.

Harris, R. Cole, ed. *Historical Atlas of Canada*. Vol. 1, *From the Beginning to 1800*. Toronto: University of Toronto Press, 1987.

Hatton, Joseph, and Moses Harvey. *Newfoundland: The Oldest British Colony*. London: Chapman and Hall, 1883.

Hay, Douglas et al. *Albion's Fatal Tree: Crime and Society in Eighteenth-Century England*. New York: Pantheon Books, 1975.

Head, C. Grant. *Eighteenth Century Newfoundland*. Toronto: McClelland and Stewart, 1976.

Hiller, James, and Peter Neary, eds. *Newfoundland in the Nineteenth and Twentieth Centuries: Essays in Interpretation*. Toronto: University of Toronto Press, 1980.

Hilton, George W. *The Truck System, Including a History of the British Truck Acts 1465–1960*. Westport, Conn.: Greenwood Press, 1960.

Hilton, Rodney, ed. *The Transition from Feudalism to Capitalism*. London: New Left Books, 1976.

Hobsbawm, Eric, and Terence Ranger, eds. *The Invention of Tradition*. Cambridge: Cambridge University Press, 1983.

Hoppit, Julian. *Risk and Failure in English Business, 1700–1800*. Cambridge: Cambridge University Press, 1987.

Hornsby, Stephen J. *Nineteenth-Century Cape Breton: A Historical Geography*. Montreal and Kingston: McGill-Queen's University Press, 1992.

Howell, Colin, and Richard Twomey, eds. *Jack Tar in History: Essays in the History of Maritime Life and Labour*. Fredericton, NB: Acadiensis Press, 1991.

Howley, Michael F. *Ecclesiastical History of Newfoundland*. Boston: Doyle & Whittle, 1888.

Innis, H.A. *The Cod Fisheries. The History of an International Economy*. Rev. ed. Toronto: University of Toronto Press, 1954.

Inwood, Kris, ed. *Farm, Factory and Fortune: New Studies in the Economic History of the Maritime Provinces*. Fredericton, NB: Acadiensis Press, 1993.

Johnson, Leo A. *History of the County of Ontario, 1615–1875*. Whitby, Ont.: Corporation of the County of Ontario, 1973.

Kealey, G.S., ed. *Class, Gender, and Region: Essays in Canadian Historical Sociology*. St John's, Nfld.: Committee on Canadian Labour History, 1988.

Knapland, Paul. *James Stephen and the British Colonial System, 1813–1847*. Madison: University of Wisconsin Press, 1953.

Knorr, Klaus E. *British Colonial Theories, 1570–1850*. Toronto: University of Toronto Press, 1944.

Kriedte, Peter, Hans Medick, and Jürgen Schlubohm, eds. *Industrialization before Industrialization: Rural Industry in the Genesis of Capitalism*. Cambridge: Cambridge University Press, 1981.

Kulikoff, Allan. *The Agrarian Origins of American Capitalism.* Charlottesville, NC: University of North Carolina Press, 1992.

Kussmaul, Ann. *Servants in Husbandry in Early Modern England.* Cambridge: Cambridge University Press, 1918.

Lahey, Raymond. *James Louis O'Donel in Newfoundland, 1784–1807.* St John's: Newfoundland Historical Society, 1984.

Laxer, Gordon. *Open for Business: The Roots of Foreign Ownership in Canada.* Toronto: Oxford University Press, 1989.

Laxer, Gordon, ed. *Perspectives on Canadian Economic Development: Class, Staples, Gender, and Elites.* Toronto: Oxford University Press, 1991.

Levine, David, ed. *Proletarianization and Family History.* London: Academic Press, 1984.

Light, Beth, and Alison Prentice, eds. *Pioneer Gentlewomen of British North America.* Toronto: New Hogtown Press, 1980.

Loo, Tina. *Making Law, Order, and Authority in British Columbia, 1821–1871.* Toronto: University of Toronto Press, 1994.

Lounsbury, Ralph G. *The British Fishery at Newfoundland, 1634–1763.* New York: Archon Books, 1969.

MacKinnon, Neil. *This Unfriendly Soil: The Loyalist Experience in Nova Scotia, 1783–1791.* Kingston and Montreal: McGill-Queens University Press, 1986.

MacNutt, W.S. *The Atlantic Provinces: The Emergence of Colonial Society.* Toronto: McClelland and Stewart, 1965.

Manning, Helen Taft. *British Colonial Government after the American Revolution 1782–1820.* New Haven, Conn.: Yale University Press, 1933.

Mannion, John J., ed. *The Peopling of Newfoundland: Essays in Historical Geography.* St John's: Institute for Social and Economic Research, Memorial University of Newfoundland, 1977.

Marx, Karl. *Grundrisse* (1857–58). Middlesex, Eng.: Penguin, 1967.

– *Capital* (1867). New York: Vintage Books, 1976.

– *Theories of Surplus-Value* (1862–63). Moscow: Progress Publishers, 1963.

Matthews, David Ralph. *Controlling Common Property: Regulating Canada's East Coast Fishery.* Toronto: University of Toronto Press, 1993.

Matthews, Keith. *Lectures on the History of Newfoundland, 1500–1830.* St John's: Breakwater, 1988.

McCalla, Douglas. *The Upper Canada Trade, 1834–1872: A Study of the Buchanan's Business.* Toronto: University of Toronto Press, 1979.

– *Planting the Province: The Economic History of Upper Canada, 1784–1870.* Toronto: University of Toronto Press, 1993.

McCallum, John. *Unequal Beginnings: Agriculture and Economic Development in Quebec and Ontario until 1870*. Toronto: University of Toronto Press, 1980.

McCann, L.D., ed. *Heartland and Hinterland*. Scarborough, Ont.: Prentice-Hall, 1987.

McCusker, John J., and Russell R. Menard. *The Economy of British North America, 1670-1789*. Chapel Hill, NC: University of North Carolina Press, 1985.

McDonald, Ian. *'To Each His Own': William Coaker and the Fishermen's Protective Union in Newfoundland Politics, 1908-1925*. Edited by J.K. Hiller. St John's: Institute of Social and Economic Research, Memorial University of Newfoundland, 1987.

McLintock, A.H. *The Establishment of Constitutional Government in Newfoundland, 1783-1832: A Study of Retarded Colonisation*. London: Longmans, Green, 1941.

McMichael, Philip. *Settlers and the Agrarian Question*. Cambridge: Cambridge University Press, 1984.

Miles, Robert. *Capitalism and Unfree Labour: Anomaly or Necessity?* London: Tavistock, 1987.

Montgomery, David. *Citizen Worker: The Experience of Workers in the United States with Democracy and the Free Market during the Nineteenth Century*. Cambridge, Mass.: Cambridge University Press, 1993.

Murphy, Terrence, and Cyril J. Byrne, eds. *Religion and Identity*. St John's, Nfld.: Jesperson Press, 1984.

Murray, Hilda. *More than 50%*. St John's: Institute for Social and Economic Research, Memorial University of Newfoundland, 1979.

Naylor, R.T. *The History of Canadian Business, 1867-1914*. 2 vols. Toronto: Lorimer, 1975.

– *Canada in the European Age, 1453-1919*. Vancouver: New Star Books, 1987.

Neary, Peter, and Patrick O'Flaherty. *By Great Waters*. Toronto: University of Toronto Press, 1974.

Neis, Barbara, and Marilyn Porter, eds. *Their Lives and Times: Women in Newfoundland and Labrador*. St John's: Creative, 1994.

Newfoundland Law Reform Commission. *Legislative History of the Judicature Act, 1791-1988*. St John's: NLRC, 1989.

Noel, S.J.R. *Politics in Newfoundland*. Toronto: University of Toronto Press, 1971.

Ommer, Rosemary. *From Outpost to Outport: A Structural Analysis of the Jersey-Gaspé Cod Fishery, 1767-1886*. Montreal and Kingston: McGill-Queen's University Press, 1991.

Ommer, Rosemary E., ed. *Merchant Credit and Labour Strategies in Historical Perspective*. Fredericton, NB: Acadiensis Press, 1990.

Orren, Karen. *Belated Feudalism: Labor, the Law, and Liberal Development in the*

United States. Cambridge: Cambridge University Press, 1991.

Palmer, Bryan D. *Working-Class Experience: Rethinking the History of Canadian Labour, 1800-1991*. Toronto: McClelland and Stewart, 1992.

Pedley, Charles. *The History of Newfoundland*. London: Longman, Green, 1867.

Pentland, H.C. *Labour and Capital in Canada*. Edited by Paul Philips. Toronto: Lorimer, 1981.

Petryshyn, J., D.M. Calman, T.S. Crossen and L. Dzubak, eds. *Victorian Cobourg*. Belleville, Ont.: Mika Studio, 1976.

Prowse, D.W. *A History of Newfoundland from the English, Colonial and Foreign Records* (1895). Belleville, Ont.: Mika Studio, 1972.

Rawlyk, George, ed. *Historical Essays on the Atlantic Provinces*. Toronto: McClelland and Stewart, 1967.

Read, Colin, and Ronald J. Stagg, eds. *The Rebellions of 1837 in Upper Canada*. Ottawa: The Champlain Society, Carleton University Press, 1985.

Rogers, J.D. *A Historical Geography of the British Colonies*. Vol. 5, Pt. 4, *Newfoundland* (1911). Oxford: Clarendon Press, 1931.

Romney, Paul. *Mr Attorney: The Attorney General for Ontario in Court, Cabinet, and Legislature, 1791-1899*. Toronto: University of Toronto Press, 1986.

Roxborough, Ian. *Theories of Underdevelopment*. London: Macmillan, 1979.

Rudé, George. *The Crowd in History: A Study of Popular Disturbances in France and England, 1730-1848*. London: Lawrence and Wishart, 1964, 1981.

- *Wilkes and Liberty*. Oxford: Oxford University Press, 1962.

Ryan, Shannon. *Fish Out of Water: The Newfoundland Saltfish Trade, 1814-1914*. St John's: Breakwater, 1986.

Sansom, Daniel, ed. *Contested Countryside: Rural Workers and Modern Society in Atlantic Canada, 1800-1950*. Fredericton, NB: Acadiensis Press, 1994.

Shaw, A.G.L., ed. *Great Britain and the Colonies, 1815-1865*. London: Methuen, 1970.

Sider, Gerald S. *Culture and Class in Anthropology and History: A Newfoundland Illustration*. Cambridge: Cambridge University Press, 1986.

Smallwood, J.R. *The New Newfoundland: An Account of the Revolutionary Developments Which Are Transforming Britain's Oldest Colony from 'The Cinderella of the Empire' into One of the Great Small Nations of the World*. New York: Macmillan, 1931.

Smallwood, J.R., ed. *Dr William Carson: The Great Newfoundland Reformer*. St John's: Newfoundland Book Publishers, 1978.

Smith, Adam. *An Inquiry into the Nature and Causes of the Wealth of Nations*, (1776). New York: Random House, 1937.

Steinfeld, Robert J. *The Invention of Free Labor: The Employment Relation in English and American Law and Culture, 1350-1870*. Chapel Hill, NC: University of North Carolina Press, 1991.

Storey, G.M., W.J. Kirwin, and J.D.A. Widdowson, eds. *Dictionary of Newfoundland English*. Toronto: University of Toronto Press, 1982.

Talbot, Thomas. *Newfoundland: Or a Letter Addressed to a Friend in Ireland in Relation to the Condition and Circumstances of the Island of Newfoundland, with an Especial View to Emigration*. London: Sampson Low, Marsten, Searle and Rivington, 1882.

Thompson, E.P. *The Making of the English Working Class*. Harmondsworth, Eng.: Penguin, 1963.

Thompson, Frederic *The French Shore Problem in Newfoundland*. Toronto: University of Toronto Press, 1961.

Tomlins, Christopher L. *Law, Labor and Ideology in the Early American Republic*. Cambridge: Cambridge University Press, 1993.

Vance, James E., Jr. *The Merchant's World*. Englewood Cliffs, NJ: Prentice-Hall, 1970.

Wallerstein, Immanuel. *The Modern World System*. Vol. 1, *Capitalist Agriculture and the Origins of the European World Economy in the Sixteenth Century*. London: Academic Press, 1974.

– *The Capitalist World-Economy*. Cambridge: Cambridge University Press, 1979.

– *The Modern World-System*. Vol. 2, *Mercantilism and the Consolidation of the European World-Economy, 1600–1750*. London: Academic Press, 1980.

– *Historical Capitalism*. London: Verso, 1983.

Ward, J.M. *Colonial Self-Government: The British Experience, 1759–1856*. Toronto: University of Toronto Press, 1976.

Williams, Glenn. *Not for Export: Toward a Political Economy of Canada's Arrested Industrialization*. Toronto: McClelland and Stewart, 1983.

Wilson, Bruce G. *The Enterprises of Robert Hamilton: A Study of Wealth and Influence in Early Upper Canada, 1776–1812*. Ottawa: Carleton University Press, 1983.

Woolf, Eric R. *Europe and the People without History*. Berkeley: University of California Press, 1982.

Wynn, Graeme. *Timber Colony: A Historical Geography of Early Nineteenth-Century New Brunswick*. Toronto: University of Toronto Press, 1981.

Young, Brian, and John A. Dickinson. *A Short History of Quebec: A Socio-Economic Perspective*. Toronto: Copp, Clark, 1988.

Young, D.M. *The Colonial Office in the Early Nineteenth Century*. London: Longmans, 1961.

II. Articles

Acheson, T.W. 'The Nature and Structure of York Commerce in the 1820s.' *Canadian Historical Review* 50 (1969): 406–28.

- 'New Brunswick Agriculture at the End of the Colonial Era: A Reassessment.' *Acadiensis* 22 (1993): 5-26.

Appleby, Joyce. 'Commercial Farming and the "Agrarian Myth" in the Early Republic.' *The Journal of American History* 68 (1982): 833-49.

Arrighi, Giovanni, and Fortunata Piselli. 'Capitalist Development in Hostile Environments: Feuds, Class Struggles, and Migrations in a Peripheral Region of Southern Italy.' *Review* 10 (1987): 649-751.

Baldwin, R.E. 'Patterns of Development in Newly Settled Regions.' *Manchester School of Economics and Social Studies* 24 (1956): 161-79.

Bernstein, Michael A., and Sean Wilentz. 'Marketing, Commerce, and Capitalism in Rural Massachusetts.' *Journal of Economic History* 44 (1984): 171-3.

Bitterman, Rusty. 'The Hierarchy of the Soil: Land and Labour in a 19th Century Cape Breton Community.' *Acadiensis* 18 (1988): 33-55.

Bradbury, Bettina. 'Pigs, Cows and Boarders: Non-Wage Forms of Survival among Montreal Families, 1861-91.' *Labour/Le Travail* 14 (fall 1984): 9-46.

Brenner, Robert. 'The Origins of Capitalist Development: A Critique Of Neo-Smithian Marxism.' *New Left Review* 104 (1977): 25-93.

Buckner, Phillip. 'Harvey, Sir John.' *Dictionary of Canadian Biography*. Vol. 8. Toronto: University of Toronto Press, 1985.

Cadigan, Sean T. 'Paternalism and Politics: Sir Francis Bond Head, the Orange Order, and the General Election of 1836.' *Canadian Historical Review* 72 (1991): 319-47.

- 'Merchant Capital, the State, and Labour in a British Colony: Servant-Master Relations and Capital Accumulation in Newfoundland's Northeast-Coast Fishery, 1775-1799.' *Journal of the Canadian Historical Association/Revue de la Société historique du Canada*, 2 (1991): 17-42.

- 'The Staple Model Reconsidered: The Case of Agricultural Policy in Northeast Newfoundland, 1785-1855.' *Acadiensis* 21 (1992): 48-71.

Clark, Christopher. 'The Household Economy, Market Exchange and the Rise of Capitalism in the Connecticut Valley, 1800-1860.' *Journal of Social History* 13 (1979): 169-90.

Crowley, John E. 'Empire versus Truck: The Official Interpretation of Debt and Labour in the Eighteenth-Century Newfoundland Fishery.' *Canadian Historical Review* 70 (1989): 311-36.

Drache, Daniel. 'The Formation and Fragmentation of the Canadian Working Class: 1820-1920.' *Studies in Political Economy* 15 (1984): 43-90.

DuPlessis, Robert S. 'The Partial Transition to World-Systems Analysis in Early Modern European History.' *Radical History Review* 39 (1987): 11-27.

English, Christopher. 'The Development of the Newfoundland Legal System to 1815.' *Acadiensis* 20 (1990): 89–119.

Greer, Allan. 'Wage Labour and the Transition to Capitalism: A Critique of Pentland.' *Labour/Le Travail* 15 (1985): 7–22.

Henretta, James A. 'Families and Farms: *Mentalité* in Pre-industrial America.' *William and Mary Quarterly* 35 (1978): 3–32.

Hiller, J.K. 'Little, Phillip Francis.' *Dictionary of Canadian Biography.* Vol. 12, *1891–1900.* Toronto: University of Toronto Press, 1990.

Johnson, Leo A. 'Independent Commodity Production: Mode of Production or Capitalist Class Formation?' *Studies in Political Economy* 6 (1981): 93–112.

Kerr, W.B. 'When Orange and Green United, 1832–9; The Alliance of Macdonell and Gowan.' *Ontario History* 34 (1942): 34–42.

Lahey, Raymond. 'Ewer (Yore), Thomas Anthony,' 'Power, John,' and 'Scallan, Thomas.' *Dictionary of Canadian Biography,* Vol. 6, *1821–1835.* Toronto: University of Toronto Press, 1987.

– 'Fleming, Michael Anthony.' *Dictionary of Canadian Biography,* Vol. 7, *1836–1850.* Toronto: University of Toronto Press, 1988.

Lewis, F., and R.M. McInnis. 'The Efficiency of the French-Canadian Farmer in the Nineteenth Century.' *Journal of Economic History* 60 (1980): 497–514.

Little, Linda. 'Collective Action in Outport Newfoundland: A Case Study from the 1830s.' *Labour/Le Travail* 26 (1990): 7–35.

Mannion, John. 'Morris, Patrick.' *Dictionary of Canadian Biography.* Vol. 7, *1836–1850.* Toronto: University of Toronto Press, 1988.

Matthews, Keith. 'Historical Fence Building: A Critique of Newfoundland Historiography.' *The Newfoundland Quarterly* 74 (1979): 21–9.

McCann, Philip. 'Culture, State Formation and the Invention of Tradition: Newfoundland, 1832–1855.' *Journal of Canadian Studies* 23 (1988): 86–103.

McNally, David. 'Staple Theory as Commodity Fetishism: Marx, Innis and Canadian Political Economy.' *Studies in Political Economy* 6 (1981): 35–63.

– 'Technological Determinism and Canadian Political Economy: Further Contributions to a Debate.' *Studies in Political Economy* 20 (1986): 161–70.

– 'Political Economy without the Working Class?' *Labour/Le Travail* 25 (1990): 217–26.

Medick, Hans. 'The Proto-industrial Family Economy: The Structural Function of Household and Family during the Transition from Peasant Society to Industrial Capitalism.' *Social History* 32 (1972): 291–316.

Mendels, Franklin F. 'Proto-industrialization: The First Phase of the Industrialization Process.' *Journal of Economic History* 32 (1972): 241–61.

Merrill, Michael. 'Cash Is Good to Eat: Self-Sufficiency and Exchange in the Rural Economy of the United States.' *Radical History Review* 3 (1977): 42–71.

O'Flaherty, Patrick. 'Lundrigan (Landergan, Landrigan, Lanergan), James.' *Dictionary of Canadian Biography.* Vol. 6, *1821–1835.* Toronto: University of Toronto Press, 1987.

– 'Carson, William.' *Dictionary of Canadian Biography,* vol. 7, *1836–1850.* Toronto: University of Toronto Press, 1988.

– 'The Seeds of Reform: Newfoundland, 1800–1818.' *Journal of Canadian Studies* 23 (1988): 39–59.

Ommer, Rosemary E. 'Merchant Credit and the Informal Economy: Newfoundland, 1919–1929.' *Historical Papers,* 1989. Ottawa: Canadian Historical Association, 1990.

Palmer, Bryan D. 'Listening to History Rather than Historians: Reflections on Working Class History.' *Studies in Political Economy* 18 (1986): 47–84.

Pastore, Ralph. 'The Collapse of the Beothuk World.' *Acadiensis* 19 (1989): 52–71.

Porter, Marilyn. '"She Was Skipper of the Shore-Crew:" Notes on the History of the Sexual Division of Labour in Newfoundland.' *Labour/Le Travail* 15 (1985): 105–23.

Rediker, Marcus. '"Good Hands, Stout Hearts, and Fast Feet": The History and Culture of Working People in Early America.' *Labour/Le Travail* 10 (1982): 123–44.

Rothenberg, Winnifred B. 'The Market and Massachusetts Farmers, 1750–1855.' *Journal of Economic History* 61 (1981): 283–314.

Sager, Eric W. 'Dependency, Underdevelopment, and the Economic History of the Atlantic Provinces.' *Acadiensis* 17 (1987): 117–36.

Senior, Hereward, and Elinor Senior. 'Boulton, Henry John.' *Dictionary of Canadian Biography.* Vol. 9, *1861–1870.* Toronto: University of Toronto Press, 1976.

Sider, Gerald S. 'Christmas Mumming and the New Year in Outport Newfoundland.' *Past and Present* 71 (1976): 102–25.

– 'The Ties That Bind: Culture and Agriculture, Property and Propriety in the Newfoundland Village Fishery.' *Social History* 5 (1980): 1–39.

Smith, Alan K. 'Where Was the Periphery?: The Wider World and the Core of the World-Economy.' *Radical History Review* 39 (1987): 547–59.

Southy, Clive. 'The Staple Thesis, Common Property, and Homesteading.' *Canadian Journal of Economics* 11 (1978): 547–59.

Thompson, E.P. 'The Moral Economy of the English Crowd in the Eighteenth Century.' *Past and Present* 50 (1971): 76–136.

Thompson, Frederic F. 'Cochrane, Sir Thomas, John.' *Dictionary of Canadian Biography.* Vol. 10, *1871–1880.* Toronto: University of Toronto Press, 1972.

Waite, P.B. 'Sir John Gaspard Le Marchant.' *Dictionary of Canadian Biography.* Vol. 10, *1871–1880.* Toronto: University of Toronto Press, 1972.

Whiteley, William H. 'Governor Hugh Palliser and the Newfoundland and
  Labrador Fishery, 1764-1768.' *Canadian Historical Review* (1969): 141-63.
- 'Palliser (Pallisser), Sir Hugh.' *Dictionary of Canadian Biography.* Vol. 4,
  1771-1800. Toronto: University of Toronto Press, 1979.
Wilson, David. 'The Irish in North America: New Perspectives.' *Acadiensis* 18
  (1988): 127-32.
Wynn, Graeme. 'Exciting a Spirit of Emulation among the 'Plodholes': Agricultural Reform in Pre-Confederation Nova Scotia.' *Acadiensis* 20 (1990):
  5-51.

III. Theses, Dissertations, Unpublished Papers

Antler, Ellen. 'Fisherman, Fisherwoman, Rural Proletariat: Capitalist Com-
  modity Production in the Newfoundland Fishery.' PhD thesis, University of
  Connecticut, 1981.
- 'Women's Work in Newfoundland Fishing Families.' Paper deposited at the
  Centre for Newfoundland Studies, Memorial University of Newfoundland,
  1976.
Antler, Steven. 'Colonial Exploitation and Economic Stagnation in Nineteenth
  Century Newfoundland.' PhD thesis, University of Connecticut, 1975.
Bittermann, Rusty. 'Escheat!: Rural Protest on Prince Edward Island,
  1832-1842.' PhD thesis, University of New Brunswick, 1991.
Budden, Geoff. 'The Role of the Newfoundland Natives Society in the Politi-
  cal Crisis of 1840-1842.' BA honours dissertation, Memorial University of
  Newfoundland, 1983.
Cadigan, Sean T. 'The Role of the Fishing Ships' Rooms Controversy in the
  Rise of a Local Bourgeoisie: St John's, Newfoundland, 1775-1812.' Paper
  presented to the Atlantic Canada Studies Conference, St John's, 1992.
Crabb, Peter. 'Agriculture in Newfoundland: A Study in Development.' PhD
  thesis, University of Hull, 1975.
Greene, John P. 'The Influence of Religion in the Politics of Newfoundland.'
  MA thesis, Memorial University of Newfoundland, 1970.
Handcock, W. Gordon. 'An Historical Geography of the Origins of English
  Settlement in Newfoundland: A Study of the Migration Process.' PhD the-
  sis, University of Birmingham, 1979.
Harris, Leslie. 'The First Nine Years of Representative Government.' MA the-
  sis, Memorial University of Newfoundland, 11959.
Jones, Frederick. 'Bishop Field, A Study in Politics and Religion in Nineteenth
  Century Newfoundland.' PhD thesis, University of Cambridge, 1971.

Kearns, William. 'The *Newfoundlander* and Daniel O'Connell's Great Repeal Year, A Response from Britain's Oldest Colony.' Paper presented to the annual meeting of the New England Conference of the American Committee for Irish Studies, Chicopee, Mass., 1986.

Little, Linda. 'Plebian Collective Action in Harbour Grace and Carbonear, Newfoundland, 1830–1840.' MA thesis, Memorial University of Newfoundland, 1984.

MacKinnon, Robert A. 'The Growth of Commercial Agriculture around St John's, 1800–1935: A Study of Local Trade in Response to Urban Demand.' MA thesis, Memorial University of Newfoundland, 1981.

Mathews, E.F.J. 'Economic History of Poole 1756–1815.' PhD thesis, University of London, 1958.

Matthews, Keith. 'History of the West Of England-Newfoundland Fishery.' PhD thesis, Oxford University, 1968.

Ryan, Shannon. 'The Newfoundland Cod Fishery in the Nineteenth Century.' MA thesis, Memorial University of Newfoundland, 1971.

Sanger, Chesley W. 'Technological and Spatial Adaptation in the Newfoundland Seal Fishery during the Nineteenth Century.' MA thesis, Memorial University of Newfoundland, 1973.

Sweeny, Robert. 'Internal Dynamics and the International Cycle: Questions of the Transition in Montreal, 1821–1828.' PhD thesis, McGill University, 1985.

Thornton, Patricia A. 'Dynamic Equilibrium: Settlement, Population and Ecology in the Strait of Belle Isle, Newfoundland, 1840–1940.' 2 vols. PhD thesis, University of Aberdeen, 1979.

Wells, Elizabeth A. 'The Struggle for Responsible Government in Newfoundland, 1846–1855.' MA thesis, Memorial University of Newfoundland, 1966.

Index

That Newfoundland has been the subject of some
of the finest historical writing over the last couple of decades
years is one of our best known secrets. For
the connoisseur of political history there is
Neary's Newf'; for those with a anthrop
bent there is a place; for the economic historian
The powerful musings of David Alexander.
And now along comes Hope & Deception, an exemplary
work of political economy.